Morris W. Hirsch

Differential Topology

Springer-Verlag
New York Heidelberg Berlin
1976

Morris W. Hirsch
University of California
Department of Mathematics
Berkeley, California 94720

AMS Subject Classification
34C40, 54Cxx, 57D05, 57D10, 57D12, 57D20, 57D55

Library of Congress Cataloging in Publication Data

Hirsch, Morris W 1933–
Differential topology.
(Graduate texts in mathematics; 33)
Bibliography: p. 209
Includes index.
1. Differential topology. I. Title.
II. Series.
QA613.6.H57 514′.7 75-26615

Printed in the United States of America.

ISBN 0-387-90148-5 Springer-Verlag New York

ISBN 3-540-90148-5 Springer-Verlag Berlin Heidelberg

Dedicated to the memory of

HENRY WHITEHEAD

Preface

This book presents some of the basic topological ideas used in studying differentiable manifolds and maps. Mathematical prerequisites have been kept to a minimum; the standard course in analysis and general topology is adequate preparation. An appendix briefly summarizes some of the background material.

In order to emphasize the geometrical and intuitive aspects of differential topology, I have avoided the use of algebraic topology, except in a few isolated places that can easily be skipped. For the same reason I make no use of differential forms or tensors.

In my view, advanced algebraic techniques like homology theory are better understood after one has seen several examples of how the raw material of geometry and analysis is distilled down to numerical invariants, such as those developed in this book: the degree of a map, the Euler number of a vector bundle, the genus of a surface, the cobordism class of a manifold, and so forth. With these as motivating examples, the use of homology and homotopy theory in topology should seem quite natural.

There are hundreds of exercises, ranging in difficulty from the routine to the unsolved. While these provide examples and further developments of the theory, they are only rarely relied on in the proofs of theorems.

Table of Contents

Chapter 6: *Morse Theory*

Chapter 7: *Cobordism*

Chapter 8: *Isotopy*

Chapter 9: *Surfaces*

Bibliography

Appendix

Index

Introduction

> Any problem which is non-linear in character, which involves more than one coordinate system or more than one variable, or where structure is initially defined in the large, is likely to require considerations of topology and group theory for its solution. In the solution of such problems classical analysis will frequently appear as an instrument in the small, integrated over the whole problem with the aid of topology or group theory.
>
> —M. Morse, *Calculus of Variations in the Large*, 1934

> La possibilité d'utiliser le modèle differential est, à mes yeux, la justification ultime de l'emploi des modèles quantitifs dans les sciences.
>
> —R. Thom, *Stabilité Structurelle et Morphogénèse*, 1972

In many branches of mathematics one finds spaces that can be described locally by n-tuples of real numbers. Such objects are called manifolds: *a manifold is a topological space which is locally homeomorphic to Euclidean n-space* $\mathbb{R}^n$. We can think of a manifold as being made of pieces of $\mathbb{R}^n$ glued together by homeomorphisms. If these homeomorphisms are chosen to be differentiable, we obtain a *differentiable manifold.* This book is concerned mainly with differentiable manifolds.

The Development of Differentiable Topology

The concept of manifold emerged gradually from the geometry and function theory of the nineteenth century. Differential geometers studied curves and surfaces in "ordinary space"; they were mainly interested in local concepts such as curvature. Function theorists took a more global point of view; they realized that invariants of a function F of several real or complex variables could be obtained from topological invariants of the sets $F^{-1}(c)$; for "most" values of c, these are manifolds.

Riemann broke new ground with the construction of what we call Riemann surfaces. These were perhaps the first abstract manifolds; that is, they were not defined as subsets of Euclidean space.

Riemann surfaces furnish a good example of how manifolds can be used to investigate global questions. The idea of a convergent power series (in one complex variable) is not difficult. This simple local concept becomes a complex global one, however, when the process of analytic continuation is introduced. The collection of all possible analytic continuations of a convergent power series has a global nature which is quite elusive. The global aspect suddenly becomes clear as soon as Riemann surfaces are introduced: the continuations fit together to form a (single valued) function on a surface. *The surface expresses the global nature of the analytic continuation process.* The problem has become geometrized.

Riemann introduced the global invariant of the *connectivity* of a surface: this meant maximal number of curves whose union does not disconnect the surface, plus one. It was known and "proved" in the 1860's that compact corientable surfaces were classified topologically by their connectivity. Strangely enough, no one in the nineteenth century saw the necessity for proving the subtle and difficult theorem that the connectivity of a compact surface is actually *finite*.

Poincaré began the topological analysis of 3-dimensional manifolds. In a series of papers on "Analysis Situs," remarkable for their originality and power, he invented many of the basic tools of algebraic topology. He also bequeathed to us the most important unsolved problem in differential topology, known as *Poincaré's conjecture*: is every simply connected compact 3-manifold, without boundary, homeomorphic to the 3-sphere?

It is interesting to note that Poincaré used purely differentiable methods at the beginning of his series of papers, but by the end he relied heavily on combinatorial techniques. For the next thirty years topologists concentrated almost exclusively on combinatorial and algebraic methods.

Although Herman Weyl had defined abstract differentiable manifolds in 1912 in his book on Riemann surfaces, it was not until Whitney's papers of 1936 and later that the concept of differentiable manifold was firmly established as an important mathematical object, having its own problems and methods.

Since Whitney's papers appeared, differential topology has undergone a rapid development. Many fruitful connections with algebraic and piecewise linear topology were found; good progress was made on such questions as embedding, immersions, and classification by homotopy equivalence or diffeomorphism. Poincaré's conjecture is still unsolved, however. In recent years techniques and results from differential topology have become important in many other fields.

The Nature of Differential Topology

In today's mathematical sciences manifolds are found in many different fields. In algebra they occur as Lie groups; in relativity as space-time; in economics as indifference surfaces; in mechanics as phase-spaces and energy surfaces. Wherever dynamical processes are studied, (hydrodynamics, population genetics, electrical circuits, etc.) manifolds are used for the "state-space," the setting for a model of the process by a differential equation or a mapping.

In most of these examples the historical development follows the *local-to-global* pattern. Lie groups, for example, were originally "local groups" having a single parametrization as a neighborhood of the origin in $\mathbb{R}^n$. Only later did global questions arise, such as the classification of compact groups. In each case the global nature of the subject became geometrized (at least partially) by the introduction of manifolds. In mechanics, for example, the differences in the possible long-term behavior of two physical systems become clear if it is known that one energy surface is a sphere and the other is a torus.

When manifolds occur "naturally" in a branch of mathematics, there is always present some extra structure: a Riemannian metric, a binary operation, a dynamical system, a conformal structure, etc. It is often this structure which is the main object of interest; the manifold is merely the setting. But the differential topologist studies the manifold itself; the extra structures are used only as tools.

The extra structure often presents fascinating local questions. In a Riemannian manifold, for instance, the curvature may vary from point to point. *But in differential topology there are no local questions.* (More precisely, they belong to calculus.) A manifold looks exactly the same at all points because it is locally Euclidean. In fact, a manifold (connected, without boundary) is homogeneous in a more exact sense: its diffeomorphism group acts transitively.

The questions which differential topology tries to answer are global; they involve the whole manifold. Some typical questions are: Can a given manifold be embedded in another one? If two manifolds are homeomorphic, are they necessarily diffeomorphic? Which manifolds are boundaries of compact manifolds? Do the topological invariants of a manifold have any special properties? Does every manifold admit a non-trivial action of some cyclic group?

Each of these questions is, of course, a shorthand request for a *theory.* The embedding question, for example, really means: define and compute diffeomorphism invariants that enable us to decide whether M embeds in N, and in how many essentially distinct ways.

If we know how to construct all possible manifolds and how to tell from "computable" invariants when two are diffeomorphic, we would be a long way toward answering any given question about manifolds. Unfortunately, such a classification theorem seems unattainable at present, except for very special classes of manifolds (such as surfaces). Therefore we must resort to more direct attacks on specific questions, devising different theories for different questions. Some of these theories, or parts of them, are presented in this book.

The Contents of This Book

The first difficulty that confronts us in analyzing manifolds is their homogeneity. A manifold has no distinguished "parts"; every point looks like every other point. How can we break it down into simpler objects?

The solution is to artificially impose on a manifold a nonhomogeneous structure of some kind which can be analyzed. The major task then is to derive intrinsic properties of the original manifold from properties of the artificial structure.

This procedure is common in many parts of mathematics. In studying vector spaces, for example, one imposes coordinates by means of a basis; the cardinality of the basis is then proved to depend only on the vector space. In algebraic topology one defines the homology groups of a polyhedron in terms of a particular triangulation, and then proves the groups to be independent of the triangulation.

Manifolds are, in fact, often studied by means of triangulations. A more natural kind of decomposition, however, consists of the level sets $f^{-1}(y)$ of a smooth map $f: M \to \mathbb{R}$, having the simplest kinds of critical points (where Df vanish). This method of analysis goes back to Poincaré and even to Möbius (1866); it received extensive development by Marston Morse and today is called *Morse theory*. Chapter 6 is devoted to the elementary aspects of Morse theory. In Chapter 9 Morse theory is used to classify compact surfaces.

A basic idea in differential topology is that of *general position* or *transversality*; this is studied in Chapter 3. Two submanifolds A, B of a manifold N are in general position if at every point of $A \cap B$ the tangent spaces of A and B span that of N. If A and B are not in general position, arbitrarily small perturbations of one of them will put them in general position. If they are in general position, they remain in it under all sufficiently small perturbations; and $A \cap B$ is then a submanifold of the "right" dimension. A map $f: M \to N$ is *transverse* to A if the graph of f and $M \times A$ are in general position in $M \times N$. This makes $f^{-1}(A)$ a submanifold of M, and the topology of $f^{-1}(A)$ reflects many properties of f. In this way an important connection between manifolds and maps is established.

Transversality is a great unifying idea in differential topology; many results, including most of those in this book, are ultimately based on transversality in one form or another.

The theory of degrees of maps, developed in Chapter 5, is based on transversality in the following way. Let $f: M \to N$ be a map between compact oriented manifolds of the same dimension, without boundary. Suppose f is transverse to a point $y \in N$; such a point is called a regular value of f. The degree of f is the "algebraic" number of points in $f^{-1}(y)$, that is, the number of such points where f preserves orientation minus the number where f reverses orientation. It turns out that this degree is independent of y and, in fact, depends only on the homotopy class of f. If $N = S^n$ then the degree is the *only* homotopy invariant. In this way we develop a bit of classical algebraic topology: the set of homotopy classes $[M, S^n]$ is naturally isomorphic to the group of integers.

The theory of fibre bundles, especially vector bundles, is one of the strongest links between algebraic and differential topology. Patterned on the tangent and normal bundles of a manifold, vector bundles are analogous to manifolds in form, but considerably simpler to analyze. Most of the deeper diffeomorphism invariants are invariants of the tangent bundle. In Chapter 4 we develop the elementary theory of vector bundles, including the classification theorem: isomorphism classes of vector bundles over M correspond naturally to homotopy classes of maps from M into a certain Grassmann manifold. This result relates homotopy theory to differential topology in a new and important way.

Further importance of vector bundles comes from the tubular neighborhood theorem: a submanifold $B \subset M$ has an essentially unique neighborhood looking like a vector bundle over B.

In 1954 René Thom proposed the equivalence relation of *cobordism*: two manifolds are cobordant if together they form the boundary of a compact manifold. The resulting set of equivalence classes in each dimension has a natural abelian group structure. In a *tour de force* of differential and algebraic topology, Thom showed that these groups coincide with certain homotopy groups, and he carried out a good deal of their calculation. The elementary aspects of Thom's theory, which is a beautiful mixture of transversality, tubular neighborhoods, and the classification of vector bundles, is presented in Chapter 7.

Of the remaining chapters, Chapter 1 introduces the basic definitions and, proves the "easy" Whitney embedding theorem: any map of a compact n-manifold into a $(2n + 1)$-manifold can be approximated by embeddings. Chapter 2 topologizes the set of maps from one manifold to another and develops approximation theorems. A key result is that for most purposes it can be assumed that every manifold is C^∞. Much of this chapter can be skipped by a reader interested chiefly in compact C^∞ manifolds. Chapter 8 is a technical chapter on isotopy, containing some frequently used methods of deforming embeddings; these results are needed for the final chapter on the classification of surfaces.

The first three chapters are fundamental to everything else in the book. Most of Chapter 6 (Morse Theory) can be read immediately after Chapter 3; while Chapter 7 (Cobordism) can be read directly after Chapter 4. The classification of surfaces, Chapter 9, uses material from all the other chapters except Chapter 7.

The more challenging exercises are starred, as are those requiring algebraic topology or other advanced topics. The few that have two stars are really too difficult to be considered exercises, but are included for the sake of the results they contain. Three-star "exercises" are problems to which I do not know the answer.

A reference to Theorem 1 of Section 2 in Chapter 3 is written 3.2.1, or as 2.1 if it appears in Chapter 3. The section is called Section 3.2. Numbers in brackets refer to the bibliography.

Acknowledgments

I am grateful to Alan Durfee for catching many errors; to Marnie McElhiney for careful typing; and to the National Science Foundation and the Miller Institute for financial support at various times while I was writing this book.

Chapter 1

Manifolds and Maps

Il faut d'abord examiner la question de la définition des variétés.

—P. Heegard, *Dissertation*, 1892

The assemblage of points on a surface is a twofold manifoldness; the assemblage of points in tri-dimensional space is a threefold manifoldness; the values of a continuous function of n arguments an n-fold manifoldness.

—G. Chrystal, *Encyclopedia Brittanica*, 1892

The introduction of numbers as coordinates . . . is an act of violence . . .

—H. Weyl, *Philosophy of Mathematics and Natural Science*, 1949

Differential topology is the study of differentiable manifolds and maps. A manifold is a topological space which locally looks like Cartesian n-space $\mathbb{R}^n$; it is built up of pieces of $\mathbb{R}^n$ glued together by homeomorphisms. If these homeomorphisms are differentiable we obtain a differentiable manifold.

The task of differential topology is the discovery and analysis of global properties of manifolds. These properties are often quite subtle. In order to study them, or even to express them, a wide variety of topological, analytic and algebraic tools have been developed. Some of these will be examined in this book.

In this chapter the basic concepts of differential topology are introduced: differentiable manifolds, submanifolds and maps, and the tangent functor. This functor assigns to each differentiable manifold M another manifold TM called its tangent bundle, and to every differentiable map $f: M \to N$ it assigns a map $Tf: TM \to TN$. In local coordinates Tf is essentially the derivative of f. Although its definition is necessarily rather complicated, the tangent functor is the key to many problems in differential topology; it reveals much of the deeper structure of manifolds.

In Section 1.3 we prove some basic theorems about submanifolds, maps and embeddings. The key ideas of regular value and transversality are introduced. The regular value theorem, which is just a global version of the implicit function theorem, is proved. It states that if $f: M \to N$ is a map then under certain conditions $f^{-1}(y)$ will be a submanifold of M. The submanifolds

$f^{-1}(y)$ and of the map f are intimately related; in this way a powerful positive feedback loop is created:

Theorems about manifolds	$\longrightarrow$ $\longleftarrow$	Theorems about maps

This interplay between manifolds and maps will be exploited in later chapters.

Also proved in Section 1.3 is the pleasant fact that every compact manifold embeds in some $\mathbb{R}^q$. Borrowing an analytic lemma from a later chapter, we then prove a version of the deeper embedding theorem of Whitney: every map of a compact n-manifold into $\mathbb{R}^{2n+1}$ can be approximated by embeddings.

Manifolds with boundary, or ∂-manifolds, are introduced in Section 1.4. These form a natural and indeed indispensable extension of the manifolds defined in Section 1.1; their presence, however, tends to complicate the mathematics. The special arguments needed to handle ∂-manifolds are usually obvious; in order to present the main ideas without interruption we shall frequently postpone or omit entirely proofs of theorems about ∂-manifolds.

At the end of the chapter a convention is stated which is designed to exclude the pathology of non-Hausdorff and nonparacompact manifolds.

Running through the chapter is an idea that pervades all of differential topology: the passage from local to global. This theme is expressed in the very definition of manifold; every statement about manifolds necessarily repeats it, explicitly or implicitly. The proof of the regular value theorem, for example, consists in pointing out the local nature of the hypothesis and conclusion, and then applying the implicit function theorem (which is itself a passage from infinitesimal to local). The compact embedding theorem pieces together local embeddings to get a global one. Whitney's embedding theorem builds on this, using, in addition, a lemma on the existence of regular values. This proof of this lemma, as will be seen in Chapter 3, is a simple globalization of a rather subtle local property of differentiable maps.

Every concept in differential topology can be analyzed in terms of this local-global polarity. Often a definition, theorem or proof becomes clearer if its various local and global aspects are kept in mind.

0. Submanifolds of $\mathbb{R}^{n+k}$

Before giving formal definitions we first discuss informally the familiar space S^n and then more general submanifolds of Euclidean space.

The *unit n-sphere* is

$$S^n = \{x \in \mathbb{R}^{n+1} : |x| = 1\},$$

where $|x| = \left(\sum_{i=1}^{n+1} x_i^2\right)^{1/2}$. We introduce local coordinates in S^n as follows.

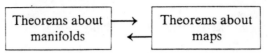

For $j = 1, \ldots, n + 1$ define open hemispheres

$$U_{2j-1} = \{x \in S^n : x_j > 0\},$$
$$U_{2j} = \{x \in S^n : x_j < 0\}.$$

For $i = 1, \ldots, 2n + 2$ define maps

$$\varphi_i : U_i \to \mathbb{R}^n,$$
$$\varphi_i(x) = (x_1, \ldots, \hat{x}_j, \ldots, x_{n+1}) \qquad \text{if } i = 2j - 1 \text{ or } 2j;$$

this means the n-tuple obtained from x by deleting the jth coordinate. Clearly φ_i maps U_i homeomorphically onto the open n-disk

$$B = \{y \in \mathbb{R}^n : |y| < 1\}.$$

It is easy to see that $\varphi_i^{-1} : B \to \mathbb{R}^{n+1}$ is analytic.

Each (φ_i, U_i) is called a "chart" for S^n; the set of all (φ_i, U_i) is an "atlas". In terms of this atlas we say a map $f : S^n \to \mathbb{R}^k$ is "differentiable of class C^r" in case each composite map

$$f \circ \varphi_i^{-1} : B \to \mathbb{R}^k$$

is C^r. If it happens that $g : S^n \to \mathbb{R}^{m+1}$ is C^r in this sense, and $g(S^n) \subset S^m$, it is natural to call $g : S^n \to S^m$ a C^r map. This definition is equivalent to the following. Let $\{(\psi_j, V_j)\}$ be an atlas for S^m, $i = 1, \ldots, q$. Then $g : S^n \to S^m$ is C^r provided each map

$$\psi_j g \varphi_i^{-1} : \varphi_i g^{-1}(V_j) \to \mathbb{R}^m$$

is C^r; this makes sense because $\varphi_i g^{-1}(V_j)$ is an open subset of $\mathbb{R}^n$.

Thus *we have extended the notion of C^r map to the unit spheres S^n, $n = 1, 2, \ldots$.* It is easy to verify that the composition of C^r maps (in this extended sense) is again C^r.

A broader class of manifolds is obtained as follows. Let $f : \mathbb{R}^{n+k} \to \mathbb{R}^k$ be a C^r map, $r \geqslant 1$, and put $M = f^{-1}(0)$. Suppose that f has rank k at every point of $f^{-1}(0)$; we call M a "regular level surface". An example is $M = S^n \subset \mathbb{R}^{n+1}$ where $f(x) = 1 - \sum_{i=1}^{n+1} x_i^2$.

Local coordinates are introduced into M as follows. Fix $p \in M$. By a linear coordinate change we can assume that the $k \times k$ matrix $\partial f_i/\partial x_j$, $1 \leqslant i, j \leqslant k$, has rank k at p. Now identify $\mathbb{R}^{n+k}$ with $\mathbb{R}^n \times \mathbb{R}^k$ and put $p = (a,b)$. According to the implicit function theorem there exist a neighborhood $U \times V$ of (a,b) in $\mathbb{R}^n \times \mathbb{R}^k$ and a C^r map $g : U \to V$, such that $g(x) = y$ if and only if $f(x,y) = 0$. Thus

$$M \cap (U \times V) = \{(x, g(x)) : x \in U\}$$
$$= \text{graph of } g.$$

Define

$$W = M \cap (U \times V),$$
$$\varphi : W \to \mathbb{R}^n,$$
$$(x, g(x)) \mapsto x \qquad (x \in U).$$

Then (φ, W) is taken as a local coordinate system on M. In terms of such coordinates we can further extend the notion of C^r map to maps between regular level surfaces.

Exactly the same constructions are made when the domain of f is taken to be an open subset of $\mathbb{R}^{n+k}$, rather than all of $\mathbb{R}^{n+k}$.

A significantly broader class of manifolds comprises those subsets M of $\mathbb{R}^{n+k}$ which *locally* are regular level surfaces of C^r maps. That is, each point of M has a neighborhood $W \subset \mathbb{R}^{n+k}$ such that

$$W \cap M = f^{-1}(0)$$

for some C^r map $f: W \to \mathbb{R}^k$ having rank k at each point $W \cap M$. Local coordinates are introduced and C^r maps are defined as before. A manifold of this type is called an "n-dimensional submanifold of $\mathbb{R}^{n+k}$".

In each of these examples it is easy to see that *the coordinate changes are* C^r. These coordinate changes are the maps

$$\varphi_j \varphi_i^{-1} : \varphi_i(U_i \cap U_j) \to \varphi_j(U_i \cap U_j)$$

where (φ_i, U_i) and (φ_j, U_j) vary over an atlas for the manifold in question. (The domain and range of $\varphi_j \varphi_i^{-1}$ are open subsets of $\mathbb{R}^m$, so that it makes sense to say that $\varphi_j \varphi_i^{-1}$ is C^r.)

This has an important implication: to verify that a map $f: M \to N$ is C^r, it suffices to check that for each point $x \in M$ there is at least one pair of charts, (φ, U) for M and (ψ, V) for N, with $x \in U$ and $f(U) \subset V$, such that the map

$$\mathbb{R}^m \supset \varphi(U) \xrightarrow{\psi f \varphi^{-1}} \psi(V) \subset \mathbb{R}^n$$

is C^r. For suppose this is true, and let $(\bar{\varphi}, \bar{U})$, $(\bar{\psi}, \bar{V})$ be any charts for M, N; we must show that $\bar{\psi} f \bar{\varphi}^{-1}$ is C^r. An arbitrary point in the domain of $\bar{\psi} f \bar{\varphi}^{-1}$ is of the form $\bar{\varphi}(x)$ where $x \in \bar{U} \cap f^{-1}(\bar{V})$. Let (φ, U), (ψ, V) be charts for M, N such that $x \in U$, $f(U) \subset V$ and $\psi f \varphi^{-1}$ is C^r. Then in a neighborhood of $\bar{\varphi}(x)$ we have

$$\bar{\psi} f \bar{\varphi}^{-1} = (\bar{\psi} \psi^{-1})(\psi f \varphi^{-1})(\varphi \bar{\varphi}^{-1}).$$

Thus $\bar{\psi} f \bar{\varphi}^{-1}$ is locally the composition of three C^r maps, so it is C^r.

Next we discuss the tangent bundle of an n-dimensional submanifold $M \subset \mathbb{R}^{n+k}$. Let $x \in M$ and let (φ, U) be a chart at x (that is, $x \in U$). Put $a = \varphi(x) \in \mathbb{R}^n$. Let $E_x \subset \mathbb{R}^{n+k}$ be the vector subspace which is the range of the linear map

$$D\varphi_a^{-1} : \mathbb{R}^n \to \mathbb{R}^{n+k}.$$

Because of the chain rule, E_x depends only on x, not on the choice of (φ, U).

The set $x \times E_x = M_x$ is called the "tangent space" to M at x. We give it the natural vector space structure inherited from E_x. Notice that $D\varphi_a^{-1}$ induces a vector space isomorphism between $\mathbb{R}^n$ and M_x.

If we associate to every $(x,y) \in M_x$ the point $x + y \in \mathbb{R}^{n+k}$, we obtain an embedding $M_x \to \mathbb{R}^{n+k}$. The image of this embedding is an affine n-plane in $\mathbb{R}^{n+k}$ passing through x. It is tangent to M in the sense that it consists of all vectors based at x which are tangents to curves in M passing through x.

If $f:M \to N$ is a C^r map (between submanifolds) and $f(x) = z$, a linear map $Tf_x:M_x \to N_z$ is defined as follows. Let (φ,U), (ψ,V) be charts for M, N at x, z. Put $\varphi(x) = a$, and define Tf_x by

$$Tf_x:(x,y) \mapsto (z, D(\psi f \varphi^{-1})_a y).$$

This is independent of the choice of (φ,U) and (ψ,V), thanks to the chain rule.

The union of all the tangent spaces of M is called the "tangent bundle" of M. The linear maps Tf_x form a map $Tf:TM \to TN$. This map plays the role of a "derivative" of the map $f:M \to N$.

By means of Tf we can extend the notion of "rank" to maps between submanifolds: the rank of f at $x \in M$ means the rank of the linear map $Tf_x:M_x \to N_y$.

The set TM is a subset of $M \times \mathbb{R}^{n+k}$, hence of $\mathbb{R}^{n+k} \times \mathbb{R}^{n+k}$. It is natural to ask whether TM is a submanifold. In fact, if (φ,U) is a chart for M, we obtain a natural chart (Φ,TU) for TM by identifying

$$TU = \{(x,y) \in TM : x \in U\}$$

and defining

$$\Phi:TU \to \mathbb{R}^n \times \mathbb{R}^n,$$
$$\Phi(x,y) = (\varphi(x),(D\varphi_a^{-1})^{-1}y).$$

These charts make TM into a C^{r-1} submanifold. The maps Tf are of class C^{r-1}.

This completes our sketch of the basic notions of *manifold*, *map* and *tangent bundle* for the special case of submanifolds of Euclidean space. We now proceed to abstract manifolds.

1. Differential Structures

A topological space M is called an n-dimensional *manifold* if it is locally homeomorphic to $\mathbb{R}^n$. That is, there is an open cover $\mathscr{U} = \{U_i\}_{i \in \Lambda}$ of M such that for each $i \in \Lambda$ there is a map $\varphi_i:U_i \to \mathbb{R}^n$ which maps U_i homeomorphically onto an open subset of $\mathbb{R}^n$. We call (φ_i,U_i) a *chart* (or *coordinate system*) with domain U_i; the set of charts $\Phi = \{\varphi_i,U_i\}_{i \in \Lambda}$ is an *atlas*.

Two charts (φ_i,U_i), (φ_j,U_j) are said to have C^r *overlap* if the *coordinate change*

$$\varphi_j\varphi_i^{-1}:\varphi_i(U_i \cap U_j) \to \varphi_j(U_i \cap U_j)$$

is of differentiability class C^r, and $\varphi_i\varphi_j^{-1}$ is also C^r. See Figure 1–1. Here r can be a natural number, ∞, or ω (meaning real analytic). This definition makes sense because $\varphi_i(U_i \cap U_j)$ and $\varphi_j(U_i \cap U_j)$ are open sets in $\mathbb{R}^n$.

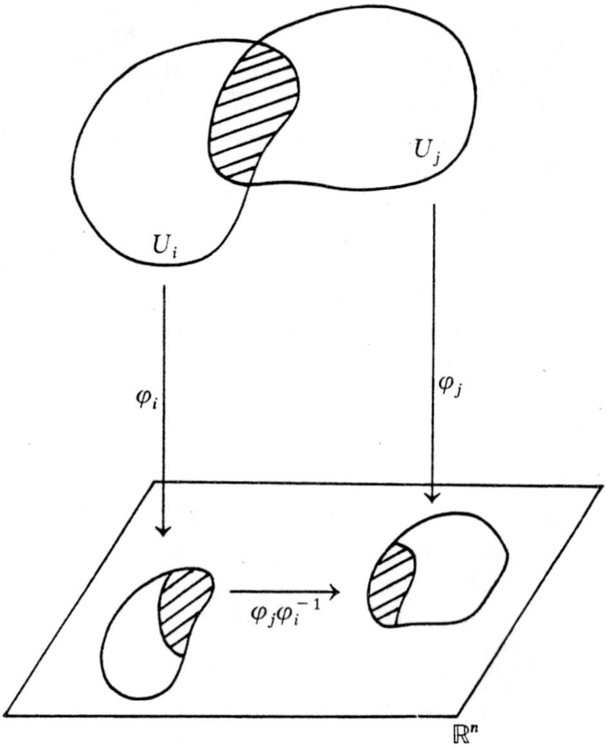

Figure 1–1. Overlapping charts.

An atlas Φ on M is called C^r if every pair of its charts has C^r overlap. In this case there is a unique maximal C^r atlas Ψ which contains Φ. In fact Ψ is the set of all charts which have C^r overlap with every chart in Φ.

A maximal C^r atlas α on M is a C^r *differential structure*; the pair (M,α) is called a *manifold of class* C^r. A manifold of class $\geqslant 1$ is called *smooth*.

To determine a C^r differential structure it suffices to give a single C^r atlas contained in it. Thus $\mathbb{R}^n$ has a unique C^r differential structure containing the identity map of $\mathbb{R}^n$. More generally every open set $U \subset \mathbb{R}^n$ has a unique C^r differential structure containing the inclusion map $U \subset \mathbb{R}^n$.

Suppose α is a C^s differential structure on M and r is an integer such that $1 \leqslant r < s$. Since α also a C^r atlas, it belongs to a unique C^r differential structure on M, obtained by adding to α all charts having C^r overlap with every chart in α. In this way every C^s manifold may be considered a C^r manifold. In Chapter 2 we shall prove the converse.

Let r be fixed until further notice; we omit the term "C^r."

If (M,Φ) and (N,Ψ) are manifolds their Cartesian product is the manifold $(M \times N,\Theta)$, where Θ is the differential structure containing all charts of the form

$$(\varphi \times \psi, U \times V); (\varphi,U) \in \Phi, (\psi,V) \in \Psi.$$

Here $\varphi \times \psi$ maps $U \times V$ into $\mathbb{R}^m \times \mathbb{R}^n$, which we identify with $\mathbb{R}^{m+n}$.

If (M,Φ) is a manifold and $W \subset M$ is an open set the *induced* differential structure on W is

$$\Phi|W = \{(\varphi,U) \in \Phi : U \subset W\}.$$

A differential structure Φ on M is often obtained by the *collation* of differential structures Φ_i on open sets U_i covering M. This means that

$$\Phi_i|U_i \cap U_j = \Phi_j|U_i \cap U_j \qquad \text{for all} \qquad i,j$$

and Φ is the unique differential structure on M containing each Φ_i as a subset.

Let M be a topological space, (N,Φ) a manifold and $h: M \to N$ a homeomorphism of M onto an open subset of N. The *induced* differential structure on M is

$$h^*\Phi = \{(\varphi h, h^{-1}U) : (\varphi,U) \in \Phi \quad \text{and} \quad U \subset h(M)\}.$$

The *n-sphere* S^n is given the C^ω differential structure defined by the atlas given in the preceding section.

Real projective n-space P^n is the C^ω manifold whose underlying space is the identification space of S^n under the antipodal map: we identify $x \in S^n$ with $-x$. If $p: S^n \to P^n$ is the natural projection, p maps each open hemisphere homeomorphically. Let $\{U_1, \ldots, U_k\}$ be a covering of S^n by open hemispheres. If we give each set $p(U_i) = V_i$ the differential structure Φ_i induced by $(p|U_i)^{-1}$, it is easy to see that Φ_i and Φ_j agree on $V_i \cap V_j$. Thus P^n is given a differential structure by collation.

More examples of manifolds are given in the exercises at the end of the section.

Some manifolds are contained in other manifolds in a natural way; thus $S^n \subset \mathbb{R}^{n+1}$. A subset A of a C^r manifold (M,Φ) is a C^r *submanifold* of (M,Φ) if for some integer $k \geq 0$, each point of A belongs to the domain of a chart $(\varphi,U) \in \Phi$ such that

$$U \cap A = \varphi^{-1}(\mathbb{R}^k)$$

where $\mathbb{R}^k \subset \mathbb{R}^n$ is the set of vectors whose last $n - k$ coordinates are 0.[1] We call such a (φ,U) a *submanifold chart* for (M,A). It is evident that if A is a submanifold of M then the maps

$$\varphi|U \cap A : U \cap A \to \mathbb{R}^k$$

form a C^r atlas for A, where (φ,U) varies over all submanifold charts. Thus A is a C^r manifold in its own right, of dimension k. The *codimension* of A is $n - k$.

[1] For $r = 0$ this is sometimes called a *locally flat* C^0 submanifold.

Let $W \subset \mathbb{R}^n$ be an open set and $f: W \to \mathbb{R}^q$ a C^r map, $1 \leqslant r \leqslant \omega$. Suppose $y \in f(W)$ is a *regular value* of f; this means that f has rank q at every point of $f^{-1}(y)$. (Therefore $q \leqslant n$.) Then the subset $f^{-1}(y)$ is a C^r submanifold of $\mathbb{R}^n$ of codimension q. This follows from the implicit function theorem, as explained in Section 1.0.

Exercises

1. The *Grassmann manifold* $G_{n,k}$ of k-dimensional linear subspaces or *k-planes* of $\mathbb{R}^n$ is given an atlas as follows. Let $E \subset \mathbb{R}^n$ be a k-plane and $E^\perp$ its orthogonal complement. Identify $\mathbb{R}^n$ with $E \times E^\perp$. Every k-plane near enough to E is the graph of a unique linear map $E \to E^\perp$. In this way a neighborhood of $E \in G_{n,k}$ is mapped homeomorphically onto an open set in the vector space of linear maps $E \to E^\perp$. This makes $G_{n,k}$ an analytic manifold of dimension $k(n-k)$.

2. *Complex projective n-space* is the manifold CP^n of (real) dimension $2n$ obtained as follows. An element of CP^n is an equivalence class $[z_0, \ldots, z_n]$ of $(n+1)$-tuples of complex numbers not all 0. The equivalence relation is: $[z_0, \ldots, z_n] = [wz_0, \ldots, wz_n]$ if w is a nonzero complex number. The topology is the natural quotient space topology. An atlas $\{\varphi_i, U_i\}$, $i = 0, \ldots, n$ is defined as follows. Let U_i be the set of equivalence classes whose i'th entry is nonzero. Map U_i into $\mathbb{C}^n$ by

$$[z_0, \ldots, z_n] \to (z_0/z_i, \ldots, \widehat{z_i/z_i}, \ldots, z_n/z_i],$$

where $\wedge$ indicates deletion. Under the natural identification of complex n-space $\mathbb{C}^n$ with $\mathbb{R}^{2n}$, these maps form a C^ω atlas on $CP(n)$.

3. *Quaternionic projective n-space* is a $4n$-dimensional manifold constructed as in Exercise 2, using quaternions instead of complex numbers.

4. The group $O(n)$ of orthogonal $n \times n$ matrices is a compact submanifold of the vector space $\mathbb{R}^{n^2}$ of all $n \times n$ matrices; its dimension is $\sum_{k=0}^{n-1} k$. The component of the identity is the subgroup $SO(n)$ of orthogonal matrices of determinant 1.

5. Let $\Phi = \{\varphi_i, U_i\}_{i \in \Lambda}$ be an atlas on an n-dimensional manifold M. Put $\varphi_i(U_i) = V_i \subset \mathbb{R}^n$, and let X be the identification space obtained from $\bigcup_{i \in \Lambda} V_i \times i$ when (x,i) is identified with $(\varphi_j \varphi_i^{-1}(x), j)$. Then X is homeomorphic to M.

6. If A is a submanifold of M, then A is a (relatively) closed submanifold of an open submanifold of M.

7. Let $G_\lambda \subset \mathbb{R} \times \mathbb{R}$ be the graph of $y = x^\lambda$, $0 \leqslant \lambda < \infty$. If $\lambda \in \mathbb{Z}$ then G_λ is a C^ω submanifold; but if $r \in \mathbb{Z}$ and $r < \lambda < r+1$ then G_λ is a submanifold which is C^r but not C^{r+1}.

8. An *atlas of class* C^r on a *set* X is sometimes defined as a collection of bijective maps from subsets of X to open subsets of $\mathbb{R}^n$ such that all coordinate changes are C^r. Given such an atlas Φ, there is a unique topology on X making Φ a C^r atlas (as defined in the text) on the space X.

9. Let C be the set of countable ordinal numbers. Let $M = C \times (-\infty, 0]$. Give M the total ordering

$$(\alpha, t) < (\alpha', t') \quad \text{if} \quad \alpha < \alpha' \quad \text{or} \quad \alpha = \alpha' \quad \text{and} \quad t < t'.$$

Endow M with the order topology. Then M is a 1-manifold which is Hausdorff but not paracompact, called the *long line*. M has a C^ω differential structure but no Riemannian metric. (See Koch and Puppe [1], Kneser and Kneser [1].)

10. Let L be the quotient space obtained from $(\mathbb{R} \times 1) \cup (\mathbb{R} \times 0)$ by identifying $(x,1)$ with $(x,0)$ if $x \neq 0$. Then L is a nonHausdorff 1-manifold, called the *line with two origins*. It has a C^ω differential structure.

***11.** (a) Let $U \subset \mathbb{R}^2$ be a nonempty open set. Suppose given a C^r $(r > 0)$ vector field on U without zeros, such that each integral curve is closed in U. Let M be the identification space obtained by collapsing each integral curve to a point. Then M is a C^r 1-manifold, which can be non-Hausdorff. [Hint: Use small intervals transverse to the integral curves to construct charts.]

****12.** A manifold is metrizable, and has a complete metric, if and only if it is paracompact and Hausdorff. A connected metrizable manifold has a countable base. But there is a connected separable Hausdorff 2-manifold which is not paracompact, of M in Exercise 7, Section 4.6.

****13.** A paracompact manifold is an absolute neighborhood retract (see Hanner [1]).

2. Differentiable Maps and the Tangent Bundle

From now on we shall frequently suppress notation for the differential structure on a manifold M.

Let M and N be C^r manifolds and $f: M \to N$ a map. A pair of charts (φ, U) for M and (ψ, V) for N is *adapted to* f if $f(U) \subset V$. In this case the map

$$\psi f \varphi^{-1}: \varphi(U) \to \psi(V)$$

is defined; we call it the *local representation* of f in the given charts, *at the point* x if $x \in U$.

The map f is called *differentiable* at x if it has a local representation at x which is differentiable. This definition makes sense since a local representation is a map between open sets in Cartesian spaces. Similarly, f is *differentiable of class* C^r if it has C^r local representations at all points.

If f is C^r then every local representation is C^r. To see this, let (φ, U) and (ψ, V) be a pair of charts adapted to f, and suppose f is C^r. To prove $\psi f \varphi^{-1}$ C^r, let $y \in \varphi(V)$ be any point; put $x = \varphi^{-1}(y)$. Let (φ_0, U_0) and (ψ_0, V_0) be an adapted pair of charts giving f the C^r local representation $\psi_0 f_0^{-1}$ at x. By replacing U_0 and V_0 by smaller open sets, if necessary, we can arrange that $U_0 \subset U$ and $V_0 \subset V$. Then

$$\psi f \varphi^{-1} = (\psi \psi_0^{-1})(\psi_0 f \varphi_0^{-1})(\varphi_0 \varphi^{-1})$$

in $\varphi(U_1)$. The first and third maps on the right are C^r since they are coordinate changes. Hence $\psi f \varphi^{-1} | \varphi(U_1)$ is the composition of C^r maps and so is C^r. This proves that $\psi f \varphi^{-1}$ is C^r in some neighborhood of every point, and so it is C^r.

Let $f:M \to N$ and $g:N \to P$ be C^r maps between C^r manifolds. It is easy to verify, using local representations, that the composition $gf:M \to P$ is also C^r. The identity map and all constant maps are C^r. There is evidently a category of C^r manifolds and C^r maps.

An isomorphism in the C^r category is called a C^r *diffeomorphism*. (If $r = 0$ this means a homeomorphism.) Explicitly, a C^r diffeomorphism $f:M \to N$ is a C^r map between C^r manifolds M and N which is a homeomorphism, and whose inverse $f^{-1}:N \to M$ is also of class C^r. If such a map exists we call M and N C^r *diffeomorphic* manifolds and write $M \approx N$. This is the basic equivalence relation of differential topology.

Lest the reader lose heart at the prospect of an infinite sequence of equivalence relations, one for each r, we hasten to point out that there is no essential difference between C^r and C^s for $1 \leqslant r < s \leqslant \infty$ (or even $s = \omega$, but that is much more difficult). In Chapter 3 we shall see that every C^r manifold is C^r diffeomorphic to a C^ω manifold, and the latter is unique up to C^ω diffeomorphism; and any C^r map can be approximated by C^ω maps.

There is, however, an unbridgeable gap between C^0 and C^1. In fact one of the most fascinating topics in differential topology began with the discoveries by Kervaire [1] and Smale [1] of compact manifolds having no differential structure whatever. (It is known that such a "nonsmoothable" manifold must have dimension at least 4; explicit examples are known in dimension 8.)

A basic task of differential topology is to find methods for deciding whether two given manifolds diffeomorphic. Of course diffeomorphic manifolds are homeomorphic, and have the same homotopy type. Therefore the diffeomorphism problem usually takes the form: what more do we need to know about two manifolds, in addition to their having the same homotopy type, to guarantee that they are diffeomorphic?

Often a differential invariant turns out to be a topological or homotopy type invariant. (The classic example is the sum of the indices of zeros of a vector field on a compact smooth manifold, which turns out to equal the Euler characteristic.) Such an invariant cannot distinguish between nondiffeomorphic manifolds which are homeomorphic. On the other hand, when a differential invariant is a homotopy invariant as well, it is easier to compute.

One of the most important differential invariants is the tangent bundle. In later chapters we will study the tangent bundle in some detail; here we merely give its definition (as a manifold) and the definition of the tangent of a map.

Let (M,Φ) be a C^{r+1} manifold, $0 \leqslant r \leqslant \omega$, where $\infty + 1 = \infty$ and $\omega + 1 = \omega$, with $\Phi = \{\varphi_i, U_i\}_{i \in \Lambda}$. Intuitively speaking, a "tangent vector" to M at $x \in M$ is simply a vector in $\mathbb{R}^n$ together with a chart which identifies each point near x with a point of $\mathbb{R}^n$.

A tangent vector should be an object independent of any particular chart, however, so we make the following definition. A *tangent vector* to M is an equivalence class $[x,i,a]$ of triples

$$(x,i,a) \in M \times \Lambda \times \mathbb{R}^n$$

under the equivalence relation:

$$[x,i,a] = [y,j,b]$$

if and only if $x = y$ and

$$D(\varphi_j\varphi_i^{-1})(\varphi_i(x))a = b.$$

In other words, the derivative of the coordinate change at $\varphi_i(x)$ sends a to b. That this is an equivalence relation follows from the rules for derivatives of compositions and inverses.

The set of all tangent vectors is TM, the *tangent bundle* of M. The map

$$p = p_M: TM \to M,$$
$$[x,i,a] \mapsto x$$

is well defined. For any subset $A \subset M$ we put $p^{-1}(A) \equiv T_AM$; also $p^{-1}(x) = M_x$ for $x \in M$. If $U \subset M$ is open then $(U,\Phi|U)$ is also a C^{r+1} manifold, and we make the harmless identification $T_UM = TU$.

For any chart $(\varphi_i,U_i) \in \Phi$ there is a well defined bijective map

$$T\varphi_i: TU_i \to \varphi_i(U_i) \times \mathbb{R}^n \subset \mathbb{R}^n \times \mathbb{R}^n,$$
$$[x,i,a] \mapsto (\varphi_i(x),a).$$

The map

$$(T\varphi_j)(T\varphi_i)^{-1}: \varphi_i(U_i \cap U_j) \times \mathbb{R}^n \to \varphi_j(U_i \times U_j) \times \mathbb{R}^n$$

is the homeomorphism

$$(y,a) \mapsto (\varphi_j\varphi_i^{-1}(y), D(\varphi_j\varphi_i^{-1})(y)a).$$

It follows that TM has a topology making each $T\varphi_i$ a homeomorphism, and this topology is unique. Moreover, since $(T\varphi_j)(T\varphi_i)^{-1}$ is a C^r diffeomorphism, the set of charts $\{T\varphi_i, TU_i\}_{i \in \Lambda}$ is a C^r atlas on TM. In this way TM is a C^r manifold. The projection map $p: TM \to M$ is C^r. The charts $(T\varphi_i, TU_i)$ are called *natural charts* on TM.

Let $x \in U_i$. The map $T\varphi_{ix}: M_x \to \mathbb{R}^n$, defined as the composition

$$M_x \subset TU_i \xrightarrow{T\varphi} \varphi_i(U_i) \times \mathbb{R}^n \to \mathbb{R}^n,$$

is a bijection; hence it induces an n-dimensional vector space structure on M_x. *This structure is independent of* i, since if $x \in U_j$,

$$(T\varphi_{jx})(T\varphi_{ix})^{-1} = D(\varphi_j\varphi_i^{-1})(\varphi_i x)$$

which is a linear automorphism of $\mathbb{R}^n$. In this way M_x becomes a vector space, the *tangent space to M at x*. Thus TM is the disjoint union of the vector spaces M_x. It is a bundle of vector spaces, or "vector bundle." This aspect of TM will be emphasized in later chapters.

The simplest kind of tangent bundle is that of an open set $W \subset \mathbb{R}^q$. In this case we identify TW with $W \times \mathbb{R}^q$ via the inclusion chart $\varphi: W \to \mathbb{R}^q$ and the corresponding natural chart on TW. The projection $TW \to W$ is just the natural projection $W \times \mathbb{R}^q \to W$. If M is a submanifold of $\mathbb{R}^3$ we can think of tangent vectors to M as arrows and M_x as a plane, as in Figure 1–2.

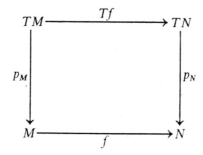

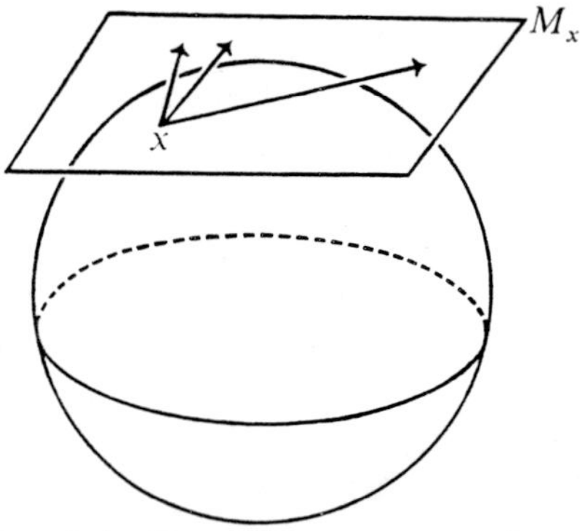

Figure 1–2. Tangent vectors to $M = S^2 \subset \mathbb{R}^3$.

Let $f: M \to N$ be a C^{r+1} map, $0 \leqslant r \leqslant \omega$. A C^r map $Tf: TM \to TN$ is defined as follows: a local representation of Tf in natural charts on TM and TN is the derivative of the corresponding local representation of f. More explicitly, let $\varphi_i: U_i \to \mathbb{R}^m$, $\psi_j: V_j \to \mathbb{R}^n$ be charts for M, N with $f(U) \subset V$. An application of the chain rule shows that the C^r map

$$\begin{aligned}(Tf)_{ij}: TU_i &\to TV_j,\\ [x,i,a] &\mapsto [f(x),j,D(\psi_j f \varphi_i^{-1})(\varphi_i x)a]\end{aligned}$$

is independent of i, j. Thus there is a well defined map $Tf: TM \to TN$ which coincides with $(Tf)_{ij}$ on TU_i.

If $f(x) = y$ then Tf maps M_x into N_y, and the restriction of Tf is a *linear* map: $T_x f: M_x \to N_y$.

In the natural charts this is just the derivative at x of the corresponding local representation of f. Thus $T_x f$ may be thought of as the derivative of f at x. Note, however, that its domain and range depend on x.

Using natural charts one sees that the diagram

$$\begin{array}{ccc} TM & \xrightarrow{\;Tf\;} & TN \\ {\scriptstyle p_M}\downarrow & & \downarrow{\scriptstyle p_N} \\ M & \xrightarrow[\;f\;]{} & N \end{array}$$

is *commutative*, that is, $f \circ p_M = p_N \circ Tf$. Likewise, if $f: M \to N$ and $g: N \to Q$ are C^{r+1} maps then the diagram

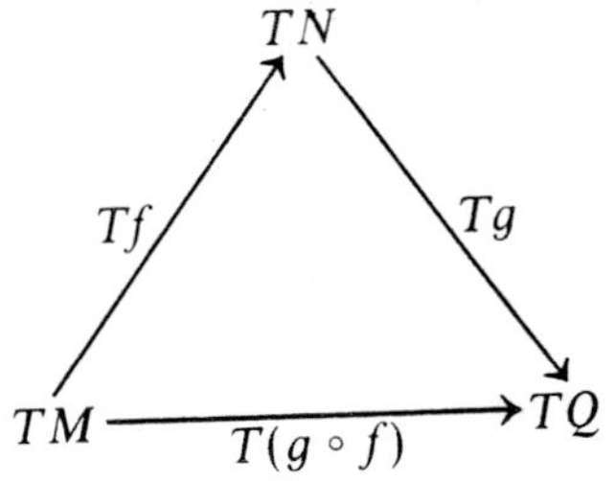

commutes; in other words

$$T(g \circ f) = (Tg) \circ (Tf).$$

And clearly

$$T\,1_M = 1_{TM}.$$

(The identity map of any space S is denoted by 1_S.) These last two properties may be summarized by saying that the assignments $M \mapsto TM$, $f \mapsto Tf$ define a covariant functor T from the category of C^{r+1} manifolds to the category of C^r manifolds.

If $M \subset N$ is a C^{r+1} submanifold, $r \geqslant 0$, let $j: M \to N$ be the inclusion map. Then $Tj: TM \to TN$ is a C^r embedding and the image of TM is a C^r submanifold of TN; this is seen by using natural charts derived from submanifold charts. Thus we identify TM with a C^r submanifold of TN.

In the special case $M \subset \mathbb{R}^q$, TM is a submanifold of $T\mathbb{R}^q = \mathbb{R}^q \times \mathbb{R}^q$.

A tangent vector to M is sometimes defined as an equivalence class of C^1 maps $f: [0,a) \to M$, where f is equivalent to $g: [0,b) \to M$ if $f(0) = g(0)$ and for some (and hence any) chart (φ_i, U_i) at $f(0)$,

$$D(\varphi_i f)(0) = D(\varphi_i g)(0).$$

To such an equivalence class we associate the tangent vector (as defined previously)

$$[f(0), i, D(\varphi_i f)(0)].$$

Conversely, to a tangent vector $[x,i,a]$ we associate the equivalence class of the C^1 map

$$f: [0,a) \to M,$$
$$f(t) = \varphi_i^{-1}(\varphi_i(x) + ta),$$

which is defined for sufficiently small $a > 0$.

These processes are inverse to each other; the two definitions of tangent vector are equivalent. But the first definition works better for manifolds with boundary.

We introduce two special notations which extend standard usage from calculus. If $J \subset \mathbb{R}$ is an interval and $f: J \to M$ is a C^1 map, for each $x \in J$

we denote by $f'(x)$ the image under Tf of the tangent vector

$$(x,1) \in T_xJ = J \times \mathbb{R}.$$

If $U \subset \mathbb{R}^n$ is open and $f: M \to U$ is C^1, for each $x \in M$ we define the linear map

$$Df_x = Df(x): M_x \to \mathbb{R}^n$$

to be the composition

$$M_x \xrightarrow{Tf} TU = U \times \mathbb{R}^n \to \mathbb{R}^n.$$

Exercises

1. Let M, N be C^r manifolds. A map $f: M \to N$ is C^r if and only if $f \circ g: W \to N$ is C^r for every C^r map $g: W \to M$.

***2.** Let M be a C^r manifold, $r \geqslant 1$, and $A \subset M$ a connected subset. Suppose that there is a C^r retraction $f: M \to A$, i.e. $f|A$ = identity. Then A is a C^r submanifold. (The converse is proved in Chapter 4.)

3. Let A, M_1, M_2 be C^r manifolds. A map $f: A \to M_1 \times M_2$, $f(x) = (f_1(x), f_2(x))$ is C^r if and only if each map $f_i: A \to M_i$ is C^r.

4. The map $G_{n,k} \to G_{n,n-k}$, $E \mapsto E^\perp$ (Exercise 1, Section 1.1) is a C^ω diffeomorphism.

5. Let $f: \mathbb{R}^n \to \mathbb{R}^k$ be any continuous map. There exists a C^ω differential structure Φ on $\mathbb{R}^n \times \mathbb{R}^k$ such that the map

$$g: \mathbb{R}^n \to (\mathbb{R}^n \times \mathbb{R}^k, \Phi),$$
$$x \mapsto (x, f(x)),$$

is a C^ω embedding.

***6.** A connected, paracompact Hausdorff 1-manifold is diffeomorphic to the circle if it is compact, and to the line if it is not compact.

7. Let Q be a positive definite quadratic form on $\mathbb{R}^n$. Then $Q^{-1}(y)$ is diffeomorphic to S^{n-1} for all $y \neq 0$.

***8.** Every nonempty starshaped open subset of $\mathbb{R}^n$ is C^∞ diffeomorphic to $\mathbb{R}^n$. ($M \subset \mathbb{R}^n$ is *starshaped* about some $x \in \mathbb{R}^n$ if it contains the entire closed interval in R^n from x to each point of M.)

9. A C^r map which is a C^1 diffeomorphism is a C^r diffeomorphism.

10. (a) The manifold $G_{3,2}$ of 2-dimensional subspaces of $\mathbb{R}^3$ is diffeomorphic to real projective 2-space P^2.

*(b) $SO(3) \approx P^3$.

*(c) The manifold of oriented 2-dimensional subspaces of $\mathbb{R}^4$ (supply the definition) is diffeomorphic to $S^2 \times S^2$.

***11.** A subset of $\mathbb{R}^2$ which is homeomorphic to S^1 is a C^0 submanifold. (This requires Schoenflies' theorem.)

12. For each $n \geqslant 0$ there is a diffeomorphism

$$(TS^n) \times \mathbb{R} \approx S^n \times \mathbb{R}^{n+1}.$$

[Hint: there are *natural* isomorphisms $T_xS^n \oplus \mathbb{R} \approx \mathbb{R}^{n+1}$.]

13. There is a natural diffeomorphism

$$T(M \times N) \approx TM \times TN.$$

14. Let $G \subset \mathbb{R} \times \mathbb{R}$ be the graph of $y = |x|^{1/3}$. Then G has a C^∞ differential structure making the inclusion $G \to \mathbb{R} \times \mathbb{R}$ a C^∞ map.

****15.** (M. Brown [1]) Let M be an n-manifold of the form $\bigcup_{k=1}^{\infty} M_k$, where each $M_k \approx \mathbb{R}^n$ and $M_k \subset M_{k+1}$. Then $M \approx \mathbb{R}^n$.

3. Embeddings and Immersions

Let $f: M \to N$ be a C^1 map (where M and N are C^r manifolds, $r \geqslant 1$). We call f *immersive* at $x \in M$ if the linear map $T_x f: M_x \to N_{f(x)}$ is injective, and *submersive* if $T_x f$ is surjective. If f is immersive at every point of M it is an *immersion*; if it is submersive at every point, f is a *submersion*.

We call $f: M \to N$ an *embedding* if f is an immersion which f maps M homeomorphically onto its image. To indicate this we may write $f: M \hookrightarrow N$.

3.1. Theorem. *Let N be a C^r manifold, $r \geqslant 1$. A subset $A \subset N$ is a C^r submanifold if and only if A is the image of a C^r embedding.*

Proof. Suppose A is a C^r submanifold. Then A has a natural C^r differential structure derived from a covering by submanifold charts. For this differential structure the inclusion of A in N is a C^r embedding.

Conversely, suppose $f: M \hookrightarrow N$ is a C^r embedding, $f(M) = A$. The property of being a C^r submanifold has *local character*, that is it is true of $A \subset M$ if and only if it is true of $A_i \subset N_i$ where $\{A_i\}$ is an open cover of A and each N_i is an open subset of M containing A_i. It is also *invariant under C^r diffeomorphisms*, that is, $A \subset N$ is a C^r submanifold if and only if $g(A) \subset N'$ is a C^r submanifold where $g: N \to N'$ is a C^r diffeomorphism (or even a C^r embedding).

To exploit local character and invariance under diffeomorphism, let $\Psi = \{\psi_i: N_i \to \mathbb{R}^n\}_{i \in \Lambda}$ be a family of charts on N which covers A. Then find an atlas $\Phi = \{\varphi_i: M_i \to \mathbb{R}^m\}_{i \in \Lambda}$ for M such that $f(M_i) \subset N_i$ (re-indexing Ψ if necessary). Since f is an embedding, Φ and Ψ can be chosen so that $f(M_i) = A \cap N_i$. By invariance it is enough to show that $\psi_i f(M_i) \subset \mathbb{R}^n$ is a C^r submanifold. Put

$$U_i = \varphi_i(M_i) \subset \mathbb{R}^m,$$
$$f_i = \psi_i f \varphi_i^{-1}: U_i \to \mathbb{R}^n.$$

Then f_i is a C^r embedding and $g_i(U_i) = \psi_i f(M_i)$. Thus we have reduced the theorem to the special case where $N = \mathbb{R}^n$, M is an open set $U \subset \mathbb{R}^m$, and $f: U \hookrightarrow \mathbb{R}^n$ is a C^r embedding. In this case a corollary of the inverse function theorem implies that there is a C^r submanifold chart for $(\mathbb{R}^n, f(U))$ at each point of $f(U)$.

QED

The theorem just proved exhibits the interplay between *local* and *global*. The statement of the theorem asserts that an object defined by local properties—a submanifold—is the same as an object defined in global terms, namely the image of an embedding. The first part of the proof just collates the local submanifold charts (restricted to A) into a differential structure on A; this makes the inclusion map of A into an embedding.

In the second part of the proof a new idea appears: the passage from *infinitesimal* to *local*. The condition that $f:M \to N$ be an immersion is an "infinitesimal" condition in that it refers only to the limiting behavior of f at each point. The inverse function theorem is a link between an infinitesimal condition and a local condition (by which we mean a statement about the behavior of f on whole neighborhoods of points).

Before stating the next theorem we make some important definitions. Let $f:M \to N$ be a C^1 map. We call $x \in M$ a *regular point* if f is submersive at x; otherwise x is a *critical point* and $f(x)$ is a *critical value*. If $y \in N$ is not a critical value it is called a *regular value*, even if y is not in $f(M)$.[2] If $y \in f(M)$ is a regular value, $f^{-1}(y)$ is called a *regular level surface*.

The following *regular value theorem* is often used to define manifolds.

3.2. Theorem. *Let $f:M \to N$ be a C^r map, $r \geqslant 1$. If $y \in f(M)$ is a regular value then $f^{-1}(y)$ is a C^r submanifold of M.*

Proof. By using local character and invariance, as in the proof of Theorem 3.1, we reduce the theorem to the case where M is an open set in $\mathbb{R}^m$ and $N = \mathbb{R}^n$. Again the theorem follows from the inverse function theorem.

QED

Theorems 3.2 and 3.1 are somewhat dual to each other under the vague dualities *immersion-submersion* and *kernel-image*, regarding $f^{-1}(y)$ as a "kernel" of f. The duality is flawed because in Theorem 3.2 the implication is only in one direction. In fact, it is not true that every submanifold is the inverse image of a regular value; see Exercise 11.

An important extension of the last result concerns a map $f:M \to N$ which is *transverse* to a submanifold $A \subset N$. This means that whenever $f(x) = y \in A$, then

$$A_y + T_x f(M_x) = N_y;$$

that is, the tangent space to N at y is spanned by the tangent space to A at y and the image of the tangent space to M at x.

3.3. Theorem. *Let $f:M \to N$ be a C^r map, $r \geqslant 1$ and $A \subset N$ a C^r submanifold. If f is transverse to A then $f^{-1}(A)$ is a C^r submanifold of M. The codimension of $f^{-1}(A)$ in M is the same as the codimension of A in N.*

[2] This is in accordance with the principle that in mathematics a *red herring* does not have to be either red or a herring.

Proof. It suffices to prove the theorem locally. Therefore we replace the pair (N,A) by $(U \times V, U \times 0)$ where $U \times V \subset \mathbb{R}^p \times \mathbb{R}^q$ is an open neighborhood of $(0,0)$. It is easy to see that the map $f: M \to U \times V$ is transverse to $U \times 0$ if and only if the composite map

$$g: M \xrightarrow{f} U \times V \xrightarrow{\pi} V$$

has 0 for a regular value (Figure 1–3). Since $f^{-1}(U \times 0) = g^{-1}(0)$ the theorem follows from Theorem 3.2.

QED

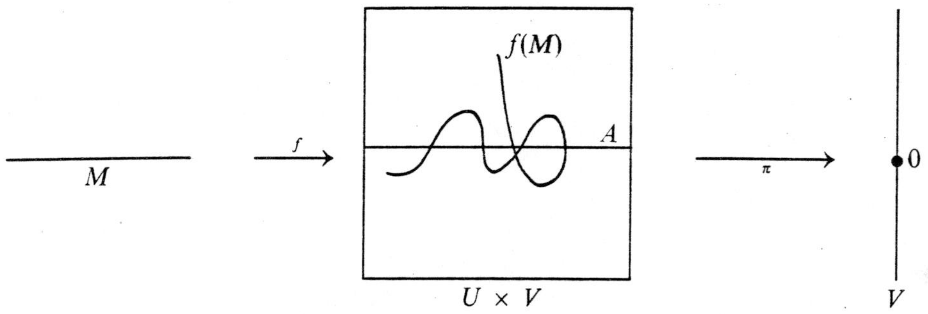

Figure 1–3. $\pi f = g$.

We shall see in the Chapter 3 that any map can be approximated by maps transverse to a given submanifold.

The next result makes the abstract notion of manifold somewhat more concrete.

3.4. Theorem. *Let M be a compact Hausdorff manifold of class C^r, $1 \leqslant r \leqslant \infty$. Then there exists a C^r embedding of M into $\mathbb{R}^q$ for some q.*

Proof. Let $n = \dim M$ be the dimension of M. Let $D^n(\rho) \subset \mathbb{R}^n$ denote the closed disk of radius ρ and center 0.[3] Since M is compact it has a finite atlas, and one easily finds an atlas $\{\varphi_i, U_i\}_{i=1}^m$ having the following two properties: for all i

$$\varphi_i(U_i) \supset D^n(2),$$

and

$$M = \bigcup \operatorname{Int} \varphi_i^{-1}(D^n(1)).$$

Let $\lambda: \mathbb{R}^n \to [0,1]$ be a C^∞ map equal to 1 on $D^n(1)$ and 0 on $\mathbb{R}^n - D^n(2)$. (Such a map is constructed in Section 2.2.) Define C^r maps

$$\lambda_i : M \to [0,1],$$

$$\lambda_i = \begin{cases} \lambda \circ \varphi_i & \text{on} \quad U_i \\ 0 & \text{on} \quad M - U_i \end{cases}$$

[3] This means that $D^n(\rho) = \{x \in \mathbb{R}^n : |x| \leq \rho\}$; the *unit disk* is $\Delta^n = D^n(1)$.

It follows that the sets

$$B_i = \lambda_i^{-1}(1) \subset U_i$$

cover M.

Define maps

$$f_i : M \to R^n,$$

$$f_i(x) = \begin{cases} \lambda_i(x)\varphi_i(x) & \text{if} \quad x \in U_i \\ 0 \quad \text{if} & x \in M - U_i \end{cases}$$

Put

$$g_i = (f_i, \lambda_i): M \to \mathbb{R}^n \times \mathbb{R} = \mathbb{R}^{n+1},$$

and

$$g = (g_1, \ldots, g_m): M \to \mathbb{R}^{n+1} \times \cdots \times \mathbb{R}^{n+1} = \mathbb{R}^{m(n+1)}.$$

Clearly g is C^r. If $x \in B_i$ then g_i, and hence g, is immersive at x, so g is an immersion. To see that g is injective, suppose $x \neq y$ with $y \in B_i$. If $x \in B_i$ then $g(x) \neq g(y)$ since $f_i|B_i = \varphi_i|B_i$. If $x \notin B_i$ then $\lambda_i(y) = 1 \neq \lambda_i(x)$, so $g(x) \neq g(y)$. Therefore g is an injective C^r immersion. Since M is compact g is an embedding.

QED

The preceding proof follows a globalization pattern that is typical in differential topology: a global construction (the embedding) is made by piecing together local objects (the charts φ_i). In this case the local embedding is implicit in the definition of manifold, but often the local construction is the more difficult part.

In most problems one runs into an "obstruction" to globalizing. If that happens, a successful theory consists of first formalizing the obstruction as a number, or other algebraic object, and then relating it to other invariants. We shall see many examples of this process.

The rest of this section is devoted to the following sharpening of Theorem 3.4, known as the "easy Whitney embedding theorem":

3.5. Theorem. *Let M be a compact Hausdorff C^r n-dimensional manifold, $2 \leqslant r \leqslant \infty$. Then there is a C^r embedding of M in $\mathbb{R}^{2n+1}$.*

Proof. By Theorem 3.4, M embeds in some $\mathbb{R}^q$. If $q = 2n + 1$ there is nothing more to prove; hence we assume $q > 2n + 1$. We may replace M by its image under an embedding. Therefore we assume that M is a C^r submanifold of $\mathbb{R}^q$. It is sufficient to prove that such an M embeds in $\mathbb{R}^{q-1}$, for repetition of the argument will eventually embed M in $\mathbb{R}^{2n+1}$.

Suppose then that $M \subset \mathbb{R}^q$, $q > 2n + 1$. Identify $\mathbb{R}^{q-1}$ with $\{x \in \mathbb{R}^q : x_q = 0\}$. If $v \in \mathbb{R}^q - \mathbb{R}^{q-1}$ denote by $f_v : \mathbb{R}^q \to \mathbb{R}^{q-1}$ the projection parallel to v. *We seek a vector v such that*

$$f_v|M : M \to \mathbb{R}^{q-1}$$

is a C^r embedding. See Figure 1–4. We limit our search to unit vectors.

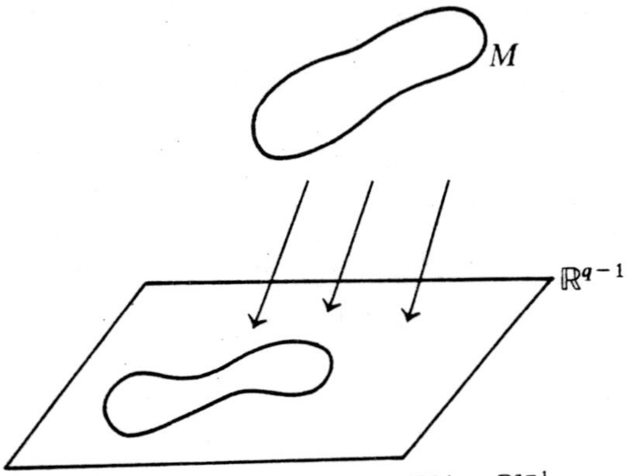

Figure 1–4. Projecting $M \subset \mathbb{R}^q$ into $\mathbb{R}^{q-1}$.

What does this require of v? For $f_v|M$ to be injective means that v is not parallel to any secant of M. That is, if x, y are any two distinct points of M, then

$$v \neq \frac{x - y}{|x - y|}. \tag{1}$$

More subtle is the requirement that $f_v|M$ be an immersion. The kernel of the linear map f_v is obviously the line through v. Therefore a tangent vector $z \in M_x$ is in the kernel of $T_x f_v$ only if z is parallel to v. We can guarantee that $f_v|M$ is an immersion by requiring, for all nonzero $z \in TM$:

$$v \neq \frac{z}{|z|} \tag{2}$$

Here z is identified with a vector in $\mathbb{R}^q$ as explained in Section 1.2; thus $|z|$ makes sense.

Condition (1) is analyzed by means of the map

$$\sigma : M \times M - \Delta \to S^{q-1},$$

$$\sigma(x,y) = \frac{x - y}{|x - y|}$$

where Δ is the *diagonal*:

$$\Delta = \{(a,b) \in M \times M : a = b\}.$$

Clearly v satisfies (1) if and only if v is *not* in the image of σ. We consider $M \times M - \Delta$ as an open submanifold of $M \times M$; the map σ is then C^r. Note that

$$\dim (M \times M - \Delta) = 2n < \dim S^{q-1}.$$

The existence of a v satisfying (1) follows from the following result:

Lemma. *Let $g : P \to Q$ be a C^1 map. If* $\dim Q > \dim P$ *then the image of g is nowhere dense in Q.*

The proof of the lemma, which involves a different set of ideas, is postponed to Chapter 3. In the case at hand $P = M \times M - \Delta$ and $Q = S^{q-1}$. Assuming the lemma, we know that every nonvoid open subset of S^{q-1} contains a point v which is not in the image of σ.

To analyze condition (2) we note that it holds for all $z \in TM$ provided it holds whenever $|z| = 1$. Let

$$T_1M = \{z \in TM : |z| = 1\}.$$

This is the *unit tangent bundle* of M. It is a C^{r-1} submanifold of TM. To see this, observe that

$$T_1M = \nu^{-1}(1)$$

where

$$\nu : TM \to \mathbb{R},$$
$$\nu(z) = |z|^2.$$

Since ν is the restriction to TM of the C^ω map

$$T\mathbb{R}^q \to \mathbb{R},$$
$$z \mapsto |z|^2,$$

it is C^{r-1}. It is clear that 1 is a regular value for ν; for if $\nu(z) = 1$ then

$$\frac{d}{dt}\nu(tz)\bigg|_{t=1} \neq 0.$$

Hence $\nu^{-1}(1)$ is a C^{r-1} submanifold by Theorem 3.2. It is easy to see that it is compact because M is compact.

Define a C^{r-1} map $\tau : T_1M \to S^{q-1}$ as follows. Identify TM with a subset of $M \times \mathbb{R}^q$; then T_1M is a subset of $M \times S^{q-1}$. Define τ to be the restriction to T_1M of the projection onto S^{q-1}. Geometrically τ is just parallel translation of unit vectors based at points of M to unit vectors based at 0.

Clearly τ is C^{r-1}. Noting that

$$\dim T_1M = 2n - 1 < \dim S^{q-1},$$

we apply the lemma to conclude that the image of τ is nowhere dense. Since T_1M is compact, it follows that the complement W of the image of τ is a dense open set in S^{q-1}. Therefore W meets $S^q \cap (\mathbb{R}^q - \mathbb{R}^{q-1})$ in a nonempty open set W_0. As we saw previously W_0 contains a vector v which is not in the image of σ. This vector v has the property that $f_v|M : M \to \mathbb{R}^q$ is an injective immersion. Since M is compact and Hausdorff, $f_v|M$ is also an embedding.

QED

There are some remarks to be made concerning the theorem just proved. It is easily converted to an *approximation theorem*: given any C^r map $g : M \to R^k$, $k \geqslant 2n + 1$, and any $\varepsilon > 0$, there is a C^r embedding $f : M \to \mathbb{R}^k$ such that $|f(x) - g(x)| < \varepsilon$ for all $x \in M$. To prove this, let $h : M \to \mathbb{R}^s$ be a

C^r embedding for some s. Then the map

$$H = g \times h: M \to \mathbb{R}^k \times \mathbb{R}^s$$

is a C^r embedding, and g is the composition of H with the projection $\pi: \mathbb{R}^k \times \mathbb{R}^s \to \mathbb{R}^k$. Identifying M with $H(M) \subset \mathbb{R}^{k+s}$, we see that it suffices to approximate π by a C^r map which restricts to an embedding of M. Now π is the composition of linear projections

$$\mathbb{R}^k \times \mathbb{R}^s \to \mathbb{R}^k \times \mathbb{R}^{s-1} \to \cdots \to \mathbb{R}^k \times \mathbb{R}^1 \to \mathbb{R}^k.$$

By induction on s it suffices to prove that if $M \subset \mathbb{R}^{k+s}$, any linear projection into $\mathbb{R}^{k+s-1}$ can be approximated by a linear projection which embeds M, provided $k + s > 2n + 1$. This is exactly what was proved.

Whitney [4] showed that Theorem 3.5 can be improved: for $n > 0$, every paracompact Hausdorff n-manifold embeds in $\mathbb{R}^{2n}$; moreover it immerses in $\mathbb{R}^{2n-1}$ if $n > 1$. However, the approximation version *cannot* be improved: if S^1 is mapped into $\mathbb{R}^2$ so that the image curve crosses itself like a figure 8, no sufficiently close approximation can be injective.

The requirement $r \geqslant 2$ in Theorem 3.5 can be weakened to $r \geqslant 1$. This follows from the result in the next chapter that every C^1 manifold has a compatible C^∞ differential structure. In fact Theorem 3.5 is true for C^0 manifolds, and even for compact metric spaces; see for example, the books by Pontryagin [2] or Hurewicz and Wallman [1]. In the other direction, Theorem 3.5 is also true for real analytic manifolds; see Chapters 2 or 4. Our proof shows that if M has a C^ω embedding in some $\mathbb{R}^q$, it has one in $\mathbb{R}^{2n+1}$.

We conclude this section with the observation that Theorem 3.5 can be improved by one dimension if we want only an immersion. For we may assume $M \subset \mathbb{R}^{2n+1}$, and can then find $v \in S^{2n} - \mathbb{R}^{2n}$ satisfying (2). Thus we see that every compact Hausdorff C^r n-manifold, $r \geqslant 2$, has a C^r immersion into $\mathbb{R}^{2n}$. In fact every C^r map $M \to \mathbb{R}^{2n}$ can be approximated by immersions.

More refined approximation theorems of this type are given in Theorems 3.2.12 and 3.2.13. More sophisticated proofs are given in Section 2.4 and at the end of Chapter 3.

Exercises

1. An injective immersion might not be an embedding, since there is an injective immersion of the line in the plane whose image is a figure 8. However, an injective immersion of a compact Hausdorff manifold is an embedding.

***2.** Let M be a connected Hausdorff noncompact C^r manifold, $r \geqslant 0$. Then there is a closed C^r embedding of the half line $[0,\infty)$ into M.

3. (a) There is an immersion of the *punctured torus* $S^1 \times S^1 - \{\text{point}\}$ in $\mathbb{R}^2$. [Hint: spread out the puncture.]

*(b) There is an immersion of the punctured n-torus, $(S^1)^n - \{\text{point}\}$, in $\mathbb{R}^n$.

4. Any product of spheres can be embedded in Cartesian space of one dimension higher.

***5.** There is no immersion of the Möbius band in the plane.

6. The line with two origins (see Exercise 10, Section 1.1) immerses in $\mathbb{R}$.

7. T_1S^2 (the unit tangent bundle of S^2) is diffeomorphic to P^3.

8. Let M be a compact C^1 manifold. Every C^1 map $M \to \mathbb{R}$ has at least two critical points.

9. Let $f:S^1 \to \mathbb{R}$ be a C^1 map and $y \in \mathbb{R}$ a regular value.
(a) $f^{-1}(y)$ has an even number of points
(b) If $f^{-1}(y)$ has $2k$ points, f has at least $2k$ critical points.
*(c) Let $g:S^2 \to \mathbb{R}$ be a C^1 map and $y \in g(S^2)$ a regular value. If $g^{-1}(y)$ has k components then f has at least $k + 1$ critical points. [Use the Jordan curve theorem.]

***10.** Every C^2 map $f: T^2 \to \mathbb{R}$ has at least 3 critical points. [$T^2 = S^1 \times S^1$ is the torus. If f has only a maximum p_+ and a minimum p_- let U be a simply connected neighborhood of p_-. Let $\varphi_t: T^2 \to T^2$, $t \in \mathbb{R}$, be the gradient flow of f. Then one can show that $T^2 - p_+ = \bigcup_{t>0} \varphi_t(U)$. This makes $T^2 - p_+$ simply connected.]

11. (a) Regarding S^1 as the equator of S^2, we obtain P^1 as a submanifold of P^2. Show that P^1 is not a regular level surface of any C^1 map on P^2. [Hint: no neighborhood of P^1 in P^2 is separated by P^1.]
(b) Generalize (a) to $P^n \subset P^{n+1}$.

12. A *surface of genus p* is a 2-dimensional manifold homeomorphic to the space obtained by removing the interiors of $2p$ disjoint 2-disks from S^2 and attaching p disjoint cyclinders to their boundaries (Figure 1–5).
*(a) For each nonnegative integer p there is a polynomial map $f_p:\mathbb{R}^3 \to \mathbb{R}$ having 0 as a regular value, such that $f_p^{-1}(0)$ is a surface of genus p. For example:

$$f_0(x,y,z) = x^2 + y^2 + z^2 - 1$$
$$f_1(x,y,z) = (x^2 + y^2 - 4)^2 + z^2 - 1$$
$$f_2(x,y,z) = [4x^2(1 - x^2) - y^2]^2 + z^2 - \tfrac{1}{4}.$$

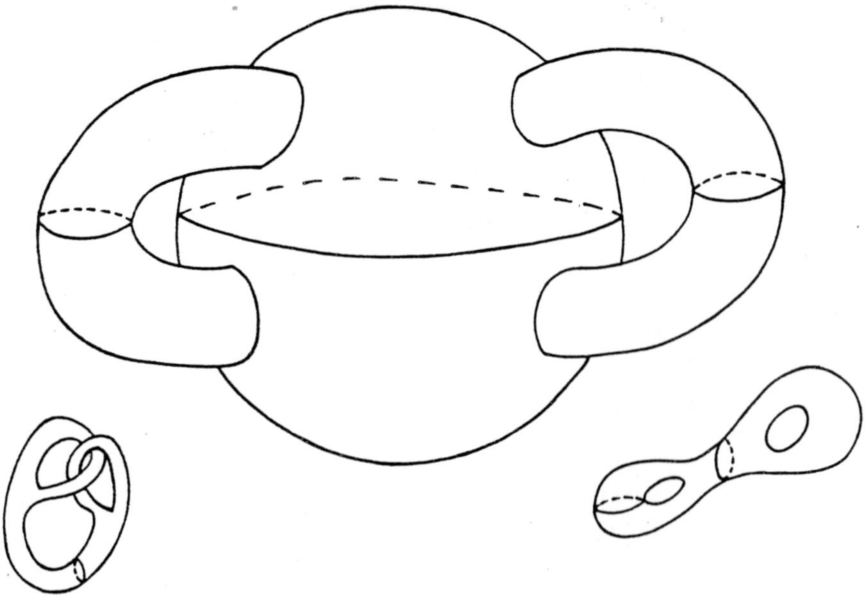

Figure 1–5. Orientable surfaces of genus $p = 2$.

[Consider functions of the form $(F(x,y))^2 + z^2 - \varepsilon^2$ where $F(x,y) = 0$ defines a closed curve in $\mathbb{R}^2$ with $p - 1$ crossings, $p \geqslant 1$.]

***(b) What is the minimal degree of f_p?

***13.** A C^1 surface of genus p has a C^1 map into $\mathbb{R}$ which has exactly 3 critical points, for all $p \geqslant 0$.

***14.** The proof of the compact embedding Theorem 3.4 can be adapted to show that every paracompact Hausdorff manifold is homeomorphic to a closed subset of a Banach space. It follows that such a manifold has a complete metric.

***15.** P^2 embeds in $\mathbb{R}^4$. [Think of P^2 as the union of a Möbius band M and a disk D. Embed M and D in $\mathbb{R}^3$ with a common boundary circle S^1; then push them out into opposite sides of $\mathbb{R}^3$ in $\mathbb{R}^4$ leaving S^1 fixed.]

16. Embeddings of P^n in S^{n+k} can be constructed as follows (Hopf [1], James [1]). Let $h:\mathbb{R}^{n+1} \times \mathbb{R}^{n+1} \to \mathbb{R}^{n+1}$ be a symmetric bilinear map such that $h(x,y) \neq 0$ if $x \neq 0$ and $y \neq 0$. Define $g:S^n \to S^{n+k}$ by $g(x) = h(x,x)/|h(x,x)|$.

(a) $g(x) = g(y)$ if and only if $x = \pm y$. [Hint:consider $h(x + \lambda y, x - \lambda y)$ if $h(x,x) = \lambda^2 h(y,y)$.]

(b) g induces an analytic embedding $P^n \to S^{n+k}$.

(c) P^n embeds in S^{2n} for all n. [Hint: let $h:\mathbb{R}^{n+1} \times \mathbb{R}^{n+1} \to \mathbb{R}^{2n+1}$,

$$h(x_0, \ldots, x_n, y_0, \ldots, y_n) = (z_0, \ldots, z_{2n})$$

where $z_k = \sum_{i+j=k} x_i y_j$.]

4. Manifolds with Boundary

Our definition of manifold excludes many objects on which differentiable maps and tangent vectors are naturally defined; the closed unit ball in $D^n \subset \mathbb{R}^n$ is an example. Many such objects are "manifolds with boundary," a concept we now explain.

A *halfspace* of $\mathbb{R}^n$, or an *n-halfspace*, is a subset of the form

$$H = \{x \in \mathbb{R}^n : \lambda(x) \geqslant 0\}$$

where $\lambda:\mathbb{R}^n \to \mathbb{R}$ is a linear map. If $\lambda \equiv 0$ then $H = \mathbb{R}^n$; otherwise H is called a *proper* halfspace. If H is proper, its *boundary* is the set $\partial H = \text{kernel}\ \lambda$; this a linear subspace of dimension $n - 1$. If $H = \mathbb{R}^n$ we set $\partial H = \varnothing$.

We now extend the definition of chart on a space M to mean a map $\varphi: U \to \mathbb{R}^n$ which maps the open set $U \subset M$ homeomorphically onto an open subset of a halfspace in $\mathbb{R}^n$. This includes all charts as defined earlier, since $\mathbb{R}^n$ is itself a halfspace; and many new charts as well. Using this definition of chart, we systematically extend the meaning of atlas, C^r atlas, C^r differential structure, and finally, C^r manifold.

Let (M,Φ) be a C^r manifold (in the new sense). Suppose $(\varphi,U) \in \Phi$ and $\varphi(U)$ is an open subset of a proper halfspace $H \subset \mathbb{R}^n$. If $x \in \varphi^{-1}(\partial H)$ we say x is a *boundary point* for the chart (φ,U). *This condition is independent of the chart.* This is the same as saying that a coordinate change cannot map an interior point of a halfspace onto a boundary point. If $r \geqslant 1$ this follows from the inverse function theorem. If $r = 0$ it follows from "invariance of domain". This is the classical and difficult topological theorem which states that a subset of $\mathbb{R}^n$ is open if it is homeomorphic to an open set; see Hurewicz and Wallman [1] for example.

The *boundary* of the manifold (M,Φ) is defined to be the set of points $x \in M$ which are boundary points for some (hence any) chart; the boundary is denoted by ∂M.

If (M,Φ) is a C^r manifold, a C^r atlas for ∂M is obtained as follows. Let $(\varphi,U) \in \Phi$ and $U \cap \partial M \neq \varnothing$. Let $H \subset \mathbb{R}^n$ be a halfspace containing $\varphi(U)$ such that $U \cap \partial M = \varphi^{-1}(\partial H)$. Let $L:\partial H \to \mathbb{R}^{n-1}$ be any linear isomorphism; then $(L\varphi, U \cap \partial M)$ is a chart on ∂M. The set of all such charts is a C^r atlas on ∂M. In this way ∂M is a C^r manifold of dimension $n - 1$.

If $\partial M \neq \varnothing$ we call M a *∂-manifold.* If $\partial M = \varnothing$ we call M a *manifold without boundary.*

The definition of C^r map between C^r manifolds is unchanged, as are the definitions of tangent vectors and the tangent bundle (if $r \geqslant 1$). The concepts of immersions, submersion, diffeomorphism and embedding go through as before.

Some care is necessary in defining "submanifold." We want, for example, a closed disk to be a submanifold of the plane. But what what about a closed disk contained in a halfspace in $\mathbb{R}^3$, whose boundary meets the boundary of the halfspace at one point? Or even worse, in a Cantor set? These are images of embeddings, and should be "submanifolds."

We first redefine C^r submanifold of $\mathbb{R}^n$ of dimension k. This is now to mean a subset $V \subset \mathbb{R}^n$ such that each point of V belongs to the domain of a chart $\psi: W \to \mathbb{R}^n$ of $\mathbb{R}^n$, such that

$$V \cap W = \psi^{-1}(H)$$

for some k-halfspace $H \subset \mathbb{R}^k \subset \mathbb{R}^n$.

Now let M be a C^r manifold, with or without boundary. A subset $A \subset M$ is a C^r submanifold if each point of A belongs to the domain of a chart $\varphi: U \to \mathbb{R}^n$ of M such that $\varphi(U \cap A)$ is a C^r submanifold (in the sense just defined) of $\mathbb{R}^n$.

An equivalent definition is this: for every $x \in A$ there is an open set $N \subset M$ containing x, a C^r embedding $g:N \to \mathbb{R}^n$ ($n = \dim M$), and a k-halfspace $H \subset \mathbb{R}^k \subset \mathbb{R}^n$ such that

$$A \cap N = g^{-1}(\mathbb{R}^k).$$

This definition of submanifold includes the old one. Theorem 3.1 is still true: a C^r submanifold, $r \geqslant 1$, is the image of a C^r embedding, and conversely.

It is useful to have a term for a submanifold $A \subset M$ whose boundary is nicely placed in ∂M. We call A a *neat* submanifold if $\partial A = A \cap \partial M$ and A is covered by charts (φ,U) of M such that

$$A \cap U = \varphi^{-1}(\mathbb{R}^m)$$

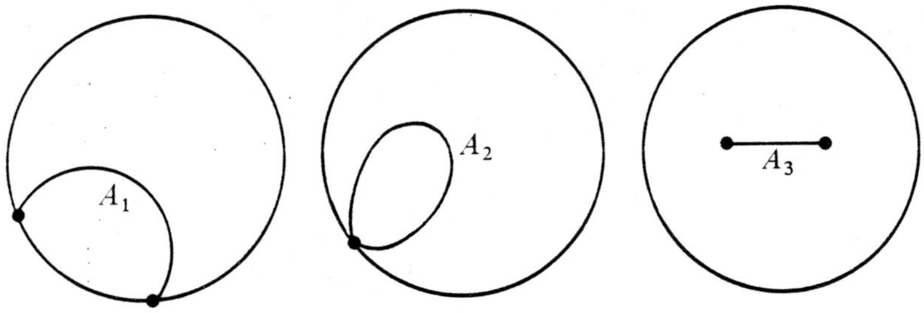

Figure 1–6. A_1 is neat; A_2 and A_3 are not.

where $m = \dim A$. (See Figure 1–6.) A *neat embedding* is one whose image is a neat submanifold.

If A is a submanifold of M and $\partial A = \varnothing$ then A is neat if and only if $A \cap \partial M = \varnothing$. In general, A is neat if and only if $\partial A = A \cap \partial M$ and (for $r \geqslant 1$) A is not tangent to ∂M at any point $x \in \partial A$; that is, $A_x \not\subset (\partial M)_x$.

The regular value theorem for ∂-manifolds takes the following form:

4.1. Theorem. *Let M be a C^r ∂-manifold and N a C^r manifold, $r \geqslant 1$. Let $f: M \to N$ be a C^r map. If $y \in N - \partial N$ a regular value for both f and $f|\partial M$, then $f^{-1}(y)$ is a neat C^r submanifold of M.*

A generalization of Theorem 3.3 to ∂-manifolds is:

4.2. Theorem. *Let $A \subset N$ be a C^r submanifold. Suppose that either A is neat, or $A \subset N - \partial N$, or $A \subset \partial N$. If $f: M \to N$ is a C^r map with both f and $f|\partial M$ transverse to A, then $f^{-1}(A)$ is a C^r submanifold and $\partial f^{-1}(A) = f^{-1}(\partial A)$.*

The proofs of Theorems 4.1 and 4.2 are left to the reader.

The embedding Theorems 3.4 and 3.5 go through with only minor changes. With some care one can prove:

4.3. Theorem. *Let M be a C^r n-dimensional manifold, $r \geqslant 1$, which is compact Hausdorff. Then there is a neat C^r embedding of M into a halfspace of $\mathbb{R}^{2n+1}$.*

The remarks following Theorem 3.5 are applicable here as well.

Exercises

1. The cartesian product of two C^0 ∂-manifolds is a ∂-manifold.

2. A C^1 map $M \to N$ takes regular points in $M - \partial M$ into $N - \partial N$.

3. Let M be the closed upper halfplane. For any C^1 map $g: R \to R$, the map $f(x, y) = y + g(x)$, from M to R, has every point of M for a regular point. Set

$$g(x) = \begin{cases} e^{-(1/x^2)} \sin(1/x) & \text{if} \quad x \neq 0 \\ 0 \quad \text{if} & x = 0. \end{cases}$$

Then $f: M \to R$ is C^∞, 0 is a regular value, but $f^{-1}(0)$ is not a manifold.

4. Let $A \subset N$ be a neat C^r submanifold, $r > 0$. Let $f:(M,\partial M) \to (N,\partial N)$ be a C^r map. Suppose every point of A [respectively ∂A] is a regular value of f [resp. $f:\partial M \to \partial N$]. Then $f^{-1}(A)$ is a neat C^r submanifold of M.

5. Let $f:M \to \mathbb{R}$ be C^r, $r > 0$. Suppose f is constant on each component of ∂M. Let a and b be regular values. Then the sets $f^{-1}(a)$, $f^{-1}[a,b]$, $f^{-1}(a,b]$, and $f^{-1}[a,\infty)$ are C^r submanifolds of M.

***6.** There is a C^∞ map $f:D^3 \to D^2$ with $0 \in D^2$ as a regular value, such that $f^{-1}(0)$ is a knotted curve (Figure 1–7).

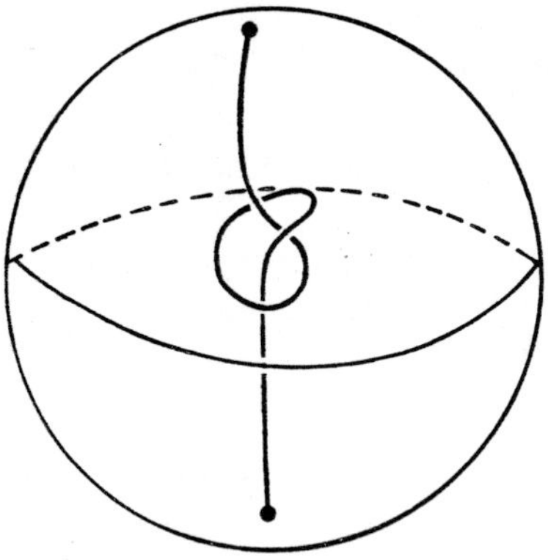

Figure 1–7.

7. (a) The *double* of a ∂-manifold M is the identification space obtained from $(M \times 0) \cup (M \times 1)$ by identifying $(x,0)$ and $(x,1)$ if $x \in \partial M$. The double is a C^0 manifold without boundary, of the same dimension, in which M is embedded.

8. If $\partial M = \varnothing$ then M is a boundary, i.e., $M = \partial N$ for some ∂-manifold N. However, if M is compact, it may be impossible to choose N compact. [Suppose M has dimension 0.]

9. A 1-dimensional connected paracompact Hausdorff ∂-manifold of class C^r, $0 \leqslant r \leqslant \infty$, is C^r diffeomorphic to either a closed, or a half open, finite interval. (This is also true, but hard, for $r = \omega$.)

10. Diffeomorphic manifolds have diffeomorphic boundaries.

11. A C^1 manifold is *orientable* if it has an atlas such that all coordinate changes have positive Jacobian determinants at every point. If M is orientable so is ∂M; but the converse can be false.

12. A subset Q of a Cartesian space $\mathbb{R}^n$ is an *orthant* if there is a linear isomorphism $L:\mathbb{R}^n \approx \mathbb{R}^{n_1} \times \cdots \times \mathbb{R}^{n_k}$ and halfspaces $H_i \subset \mathbb{R}^{n_i}$ such that $L(Q) = H_1 \times \cdots \times H_k$. There is a category of "C^r manifolds with convex corners" whose charts are homeomorphisms onto open subsets of orthants. This category contains all C^r manifolds, with and without boundaries, and is *closed under Cartesian product*.

5. A Convention

Manifolds that are not paracompact are amusing, but they never occur naturally. What is perhaps worse, it is difficult to prove anything about them. Non-Hausdorff manifolds occasionally turn up (see Exercise 11,

Section 1.1) but again it is hard to prove anything interesting. It is convenient to deal only with manifolds having a countable number of components. We therefore adopt the following convention:

All manifolds appearing henceforth are assumed to be paracompact, with a countable base, and all spaces Hausdorff, unless there is an explicit statement to the contrary.

Of course any space or manifold we construct must be shown to have these properties; the proof is usually trivial.

Chapter 2

Function Spaces

The statement sometimes made, that there exist only analytic functions in nature, is in my opinion absurd.

—F. Klein, *Lectures on Mathematics*, 1893

Many problems in differential topology can be rephrased as questions about function spaces; this often leads to new insights and greater unity. For example, in Chapter 1 we constructed "by hand" an embedding of any compact manifold M in some $\mathbb{R}^q$; in this chapter we shall exploit the topology of a space of maps of M to N to prove that any map $M \to N$ can be approximated by embeddings if $\dim N < 2 \dim M$.

The most useful topology on the set $C^r(M,N)$ of C^r maps from M to N is the strong topology. Roughly speaking, a neighborhood of f in the strong topology consists of all maps g which are close to f together with their derivatives of order r. The degree of closeness is specified by arbitrary positive numbers controlling the closeness of derivatives of local representations of f and g.

The weak (or "C^r compact-open") topology on $C^r(M,N)$ controls the closeness of maps only over compact sets. When M is compact it is the same as the strong topology.

Section 2.3 briefly indicates the changes needed to extend the approximation theorems to ∂-manifolds and manifold pairs.

In Section 2.4 jets are defined and used to give an indirect definition of the weak and strong topologies. The density of embeddings is proved again by exploiting the Baire property. The last section discusses, without proofs, various results on analytic approximations.

As deeper approximation and globalization techniques are developed they are used to improve the Whitney embedding and immersion theorems of the preceding chapter. Thus the density of immersions and embeddings is re-examined in Sections 2.1 and 2.2 and again in Section 2.4. The final form of the density of embeddings is Theorem 2.13.

1. The Weak and Strong Topologies on $C^r(M,N)$

If M and N are C^r manifolds, $C^r(M,N)$ denotes the set of C^r maps from M to N. At first we assume r is finite.

The *weak* or "compact-open C^r" topology on $C^r(M,N)$ is generated by the sets defined as follows. Let $f \in C^r(M,N)$. Let (φ,U), (ψ,V) be charts on M, N; let $K \subset U$ be a compact set such that $f(K) \subset V$; let $0 < \varepsilon \leqslant \infty$. Define a *weak subbasic neighborhood*

$$\mathcal{N}^r(f;(\varphi,U),(\psi,V),K,\varepsilon) \tag{1}$$

to be the set of C^r maps $g:M \to N$ such that $g(K) \subset V$ and

$$\|D^k(\psi f \varphi^{-1})(x) - D^k(\psi g \varphi^{-1})(x)\| < \varepsilon$$

for all $x \in \varphi(K)$, $k = 0, \ldots, r$. This means that the local representations of f and g, together with their first k derivatives, are within ε at each point of K.

The *weak topology* on $C^r(M,N)$ is generated by these sets (1); it defines the topological space $C^r_W(M,N)$. A neighborhood of f is thus any set containing the intersection of a finite number of sets of type (1).

If the proof of the easy Whitney embedding theorem (Theorem 1.3.5) is reexamined one sees that it proves the following approximation result:

1.0. Proposition. *Let M be a compact C^r manifold, $2 \leqslant r \leqslant \infty$. Then embeddings are dense in $C^r_W(M,\mathbb{R}^q)$ if $q > 2 \dim M$, while immersions are dense if $q \geqslant 2 \dim M$.*

It can be shown that C^r_W has very nice features: for example it has a complete metric and a countable base; if M is compact it is locally contractible, and $C^r_W(M,\mathbb{R}^m)$ is a Banach space.

If M is not compact the weak topology does not control the behavior of a map "at infinity" very well. For this purpose the *strong* topology is useful. (This topology is also called the *fine* or *Whitney* topology) A base consists of sets of the following type. Let $\Phi = \{\varphi_i, U_i\}_{i \in \Lambda}$ be a *locally finite* set of charts on M; this means that every point of M has a neighborhood which meets U_i for only a finite number of i. Let $K = \{K_i\}_{i \in \Lambda}$ be a family of compact subsets of M, $K_i \subset U_i$. Let $\Psi = \{\psi_i, V_i\}_{i \in \Lambda}$ be a family of charts on N, and $\varepsilon = \{\varepsilon_i\}_{i \in \Lambda}$ a family of positive numbers. If $f \in C^r(M,N)$ takes each K_i into V_i, define a *strong basic neighborhood*

$$\mathcal{N}^r(f;\Phi,\Psi,K,\varepsilon) \tag{2}$$

to be the set of C^r maps $g:M \to N$ such that for all $i \in \Lambda$, $g(K_i) \subset V_i$ and

$$\|D^k(\psi_i f \varphi_i^{-1})(x) - D^k(\psi_i g \varphi_i^{-1})(x)\| < \varepsilon_i$$

for all $x \in \varphi_i(K_i)$, $k = 0, \ldots, r$. The strong topology has all possible sets of this form for a base.

It must of course be verified these sets (2), as f, Φ, Ψ, K, ε vary, actually form a base for a topology. We leave this to the reader as an exercise; it also follows from alternative description of the topology given in Section 4.

The topological space $C^r_S(M,N)$ resulting from the strong topology is the same as $C^r_W(M,N)$ if M is compact. If M is not compact, however, and N has positive dimension, it is an extremely large topology: it is not metrizable and in fact does not have a countable base at any point; and it has uncountably many components. It does have one saving grace, however: the category theorem of Baire is valid in $C_S^r(M,N)$, as will be proved in Section 2.4.

We now define the spaces $C_W^\infty(M,N)$ and $C_S^\infty(M,N)$. The weak topology on $C^\infty(M,N)$ is simply the union of the topologies induced by the inclusion maps $C^\infty(M,N) \to C_W^r(M,N)$ for r finite, while the strong topology on $C^\infty(M,N)$ is the union of the topologies induced by $C^\infty(M,N) \to C_S^r(M,N)$. We give $C^\omega(M,N)$ the weak and strong topologies induced from $C^\infty(M,N)$.

The strong topology is very convenient for differential topology in that many important subsets are open. For example:

1.1. Theorem. *The set* $\mathrm{Imm}^r(M,N)$ *of* C^r *immersions is open in* $C_S^r(M,N)$, $r \geqslant 1$.

Proof. Since

$$\mathrm{Imm}^r(M,N) = \mathrm{Imm}^1(M,N) \cap C^r(M,N)$$

it suffices to prove this for $r = 1$. If $f: M \to N$ is a C^1 immersion one can choose a neighborhood $\mathcal{N}^1(f; \Phi,\Psi,K,\varepsilon)$ as follows. Let $\Psi^0 = \{\psi_\beta, V_\beta\}_{\beta \in B}$ be any atlas for N. Pick an atlas $\Phi = \{\varphi_i, U_i\}_{i \in \Lambda}$ for M so that each U_i has compact closure, and for each $i \in \Lambda$ there exists $\beta(i) \in B$ such that $f(U_i) \subset V_{\beta(i)}$. Put $V_{\beta(i)} = V_i$, $\psi_{\beta(i)} = \psi_i$, and $\Psi = \{\psi_i, V_i\}_{i \in \Lambda}$. Let $K = \{K_i\}_{i \in \Lambda}$ be a compact cover of M with $K_i \subset U_i$.

The set

$$A_i = \{D(\psi_i f \varphi_i^{-1})(x) | x \in \varphi_i(K_i)\}$$

is a compact set of injective linear maps from $\mathbb{R}^m$ to $\mathbb{R}^n$. Since the set of all injective linear maps is open in the vector space $L(\mathbb{R}^m,\mathbb{R}^n)$ of all linear maps $\mathbb{R}^m \to \mathbb{R}^n$, there exists $\varepsilon_i > 0$ such that $T \in L(\mathbb{R}^m,\mathbb{R}^n)$ is injective if $\|T - S\| < \varepsilon_i$ and $S \in A_i$. Set $\varepsilon = \{\varepsilon_i\}$. It follows that every element of $\mathcal{N}^1(f; \Phi,\Psi,K,\varepsilon)$ is an immersion.

QED

A similar argument, which we leave to the reader, proves:

1.2. Theorem. *The set of submersions is open in* $C_S^r(M,N)$, $1 \leqslant r \leqslant \infty$.

Our next goal is the openness of the set of embeddings. We shall need the following fact:

1.3. Lemma. *Let* $U \subset \mathbb{R}^m$ *be an open set and* $W \subset U$ *an open set with compact closure* $\overline{W} \subset U$. *Let* $f: U \to \mathbb{R}^n$ *be a* C^1 *embedding. There exists* $\varepsilon > 0$ *such that if* $g: U \to \mathbb{R}^n$ *is* C^1 *and*

$$\|Dg(x) - Df(x)\| < \varepsilon$$

for all $x \in W$ *then* $g|W$ *is an embedding.*

Proof. By Theorem 1.1 (or rather, its proof) and compactness of $\overline{W}$, there exists $\varepsilon_0 > 0$ so small that if $g \in C^1(U,\mathbb{R}^n)$ and $\|Dg(x) - Df(x)\| < \varepsilon_0$

for all $x \in W$ then $g|W$ is an immersion. Therefore if the lemma is false there is a sequence $g_n \in C^1(U,\mathbb{R}^n)$ such that

$$\|Dg_n(x) - Df(x)\| \to 0$$

and

$$(3) \qquad |g_n(x) - f(x)| \to 0$$

uniformly on W, while for each n there exist distinct points a_n, b_n in W with $g_n(a_n) = g_n(b_n)$. By compactness of $\overline{W}$ we may assume $a_n \to a \in U$, $b_n \to b \in U$ as $n \to \infty$. Then $f(a) = f(b)$ by (3), so $a = b$. Choosing subsequences if necessary we may assume that the sequence of unit vectors

$$v_n = \frac{a_n - b_n}{|a_n - b_n|}$$

converges to a unit vector $v \in S^{m-1}$. By uniformity of Taylor expansion we have

$$|g_n(a_n) - g_n(b_n) - Df(b_n)(a_n - b_n)|/|a_n - b_n| \to 0.$$

Hence $Df(b_n)v_n \to 0$. But this sequence also goes to $Df(b)v$, which therefore is 0. This contradicts the assumption that f is an immersion.

QED

We can now prove:

1.4. Theorem. *The set* $\mathrm{Emb}^r(M,N)$ *of* C^r *embeddings of* M *in* N *is open in* $C^r_S(M,N)$, $r \geq 1$.

Proof. It suffices to take $r = 1$. Let $f \in \mathrm{Emb}^r(M,N)$. By using the preceding lemma we can find the following objects:

a locally finite atlas $\Phi = \{\varphi_i, U_i\}_{i \in \Lambda}$ of M;
a set $\Psi = \{\psi_i, V_i\}_{i \in \Lambda}$ of charts for N with $f(U_i) \subset V_i$;
a family of compact sets $K_i \subset U_i$ whose interiors W_i cover M;
numbers $\varepsilon_i > 0$ such that if

$$g \in \mathcal{N}_0 = \mathcal{N}^r(f; \Phi, \Psi, K, \varepsilon)$$

then $g(U_i) \subset V_i$ and $g|W_i$ is a C^r embedding.

Since f is an embedding for each $i \in \Lambda$ there exist disjoint open sets A_i, B_i in N such that $f(K_i) \subset A_i$ and $f(M - U_i) \subset B_i$. One can find a neighborhood $\mathcal{N}_1$ of f in $C^r_S(M,N)$ (in fact, in $C^0_S(M,N)$) such that if $g \in \mathcal{N}_1$ then

$$g(K_i) \subset A_i,$$
$$g(M - U_i) \subset B_i.$$

We show that every $g \in \mathcal{N}_0 \cap \mathcal{N}_1$ is an embedding. By the choice of $\mathcal{N}_0$, g is an immersion. To see that g is injective suppose x, y are distinct points of M with $x \in K_i$. If $y \in U_i$ then $g(x) \neq g(y)$ since $g|U_i$ is injective;

while if $y \in M - U_i$ then $g(x) \in A_i$ and $g(y) \in B_i$, so again $g(x) \neq g(y)$. To see that $g: M \to g(M)$ is a homeomorphism, it suffices to show that if y_n is a sequence in M such that $g(y_n) \to g(x)$ then $y_n \to x$. If $x \in K_i$ then $g(x) \in A_i$; hence only a finite number of the $g(y_n)$ can be in B_i, so all but a finite number of y_n are in U_i. Since $g|U: U \to g(U)$ is a homeomorphism it follows that $y_n \to x$.

QED

A map f is *proper* if f^{-1} takes compact sets to compact sets.

1.5. Theorem. *The set* $\mathrm{Prop}^r(M,N)$ *of proper* C^r *maps* $M \to N$ *is open in* $C^r_S(M,N)$, $r \geq 0$.

Proof. For any map $f: M \to N$ there is a compact cover $\{K_i\}_{i \in \Lambda}$ of M and an open cover $\mathscr{V} = \{V_i\}_{i \in \Lambda}$ of N with $f(K_i) \subset V_i$. If f is proper $\mathscr{V}$ can be chosen *locally finite*. There is a neighborhood $\mathscr{N}$ of f such that if $g \in \mathscr{N}$ then $g(K_i) \subset V_i$ for all i. To see that such a g is proper, let $L \subset N$ be compact. Then L meets only a finite number of V_i. Hence $g^{-1}(L)$ is a closed subset of M which is covered by finitely many of the compact sets K_i; therefore $g^{-1}(L)$ is compact.

QED

Since an embedding $f: M \to N$ is proper if and only if $f(M)$ is closed in N, we obtain:

1.6. Corollary. *The set of closed embeddings is open in* $C^r_S(M,N)$, $r \geq 1$.

Let $\mathrm{Diff}^r(M,N)$ denote the set of C^r diffeomorphisms from M onto N.

1.6. Theorem. *If* M *and* N *are* C^r *manifolds without boundary then* $\mathrm{Diff}^r(M,N)$ *is open in* $C^r_S(M,N)$, $r \geq 1$.

Proof. A diffeomorphism induces a bijective correspondence from components of M to components of N. Such a map has a neighborhood of maps inducing the same correspondence. Therefore we may assume M and N connected.

A diffeomorphism is simultaneously an embedding, a submersion and a proper map. Conversely any map g between connected manifolds with these three properties is a diffeomorphism. For the image of a submersion is open (by the inverse function theorem) and the image of a proper map is closed; so g is a surjective embedding, which is a diffeomorphism. Thus $\mathrm{Diff}^r(M,N)$ is the intersection of three open subsets of $C^r_S(M,N)$.

QED

For ∂-manifolds Theorem 1.6 is false. But one can show that $\mathrm{Diff}^r(M,N)$ is open in the subspace

$$C^r_S(M,\partial M; N,\partial N) = \{f \in C^r_S(M,N): f(\partial M) \subset \partial N\}.$$

Theorem 1.6 is false for $r = 0$: the set of homeomorphisms is not open in $C^0_S(M,N)$ (unless it is empty or $\dim M = 0$). There is, however, the following result:

1.7. Theorem. *Let M and N be manifolds without boundary and $f: M \to N$ a homeomorphism. Then f has a neighborhood of surjective maps in $C^0_S(M,N)$.*

Proof. Let g be near f; then $g^{-1}f$ is near the identity map of M. Hence it suffices to take $M = N$ and $f = 1_M$.

Let $\{\varphi_i, U_i\}$ be a locally finite cover of M by charts such that $\varphi_i(U_i) \supset D^n$, the closed unit ball in $\mathbb{R}^n$, and $M = \cup \varphi_i^{-1}(D^n)$. For each i let $B_i \subset \varphi_i(U_i)$ be a slightly larger closed ball, $0 \in D^n \subset \operatorname{Int} B_i$. It suffices to find $\varepsilon_i > 0$ such that if $h_i: B_i \to \mathbb{R}^n$ is a continuous map with $|h_i(x) - x| < \varepsilon_i$ for all i then $h(B_i) \supset D^n$. For if this is true then the set of $g: M \to M$ satisfying, for all i,

$$g\varphi_i^{-1}(B_i) \subset U_i,$$

and

$$|\varphi_i g \varphi_i^{-1}(x) - x| < \varepsilon_i, \qquad \text{all} \qquad x \in B_i,$$

will consist of surjective maps (put $\varphi_i g \varphi_i^{-1} = h_i$).

Let $\varepsilon_i > 0$ be so small that for any $z \in D^n$, $x \in \partial B_i$ and $y \in \mathbb{R}^n$ with $|x - y| < \varepsilon_i$, it is true that the ray issuing from z through y intersects ∂B_i at a point u such that $|u - x| < \operatorname{diam} B_i$ (Figure 2–1).

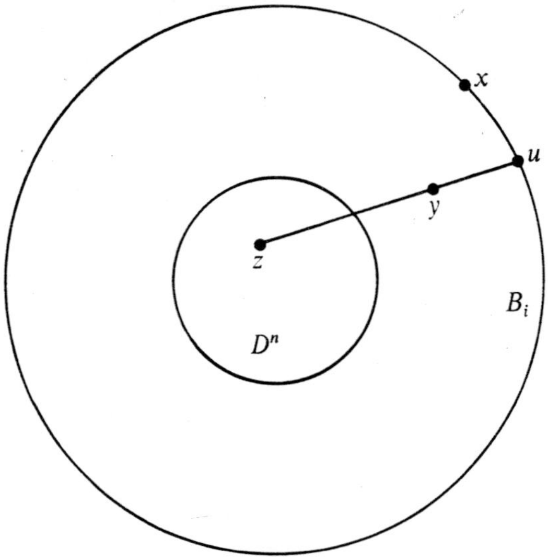

Figure 2–1.

Now suppose $h: B_i \to \mathbb{R}^n$ and $|h(x) - x| < \varepsilon_i$. Suppose $z \in D^n - h(B_i)$. Define a map $H: B_i \to \partial B_i$ by sending $x \in B_i$ to the intersection of ∂B_i with the ray from z through $h(x)$. The choice of ε_i ensures that for $x \in \partial B_i$, $H(x) \neq -x$. Then $H|\partial B_i: \partial B_i \to \partial B_i$ is homotopic to the identity; the homotopy moves $H(x)$ along the shorter great circle path from $H(x)$ to x (Figure 2–2). A classical theorem of topology, however, says that no map of an $n - 1$ sphere to itself which extends to a map of the $n - 1$ ball to the sphere can be homotopic to the identity. (This will be proved by Theorem 3.1.4). This contradiction shows that $D^n \subset h(B_i)$, proving Theorem 1.7.

QED

For ∂-manifolds one can show that any homeomorphism $h: M \to N$ has a neighborhood of surjections in $C^0_S(M,\partial M; N,\partial N)$. In fact this follows from Theorem 1.7 by extending h to a homeomorphism between the doubles of M and N.

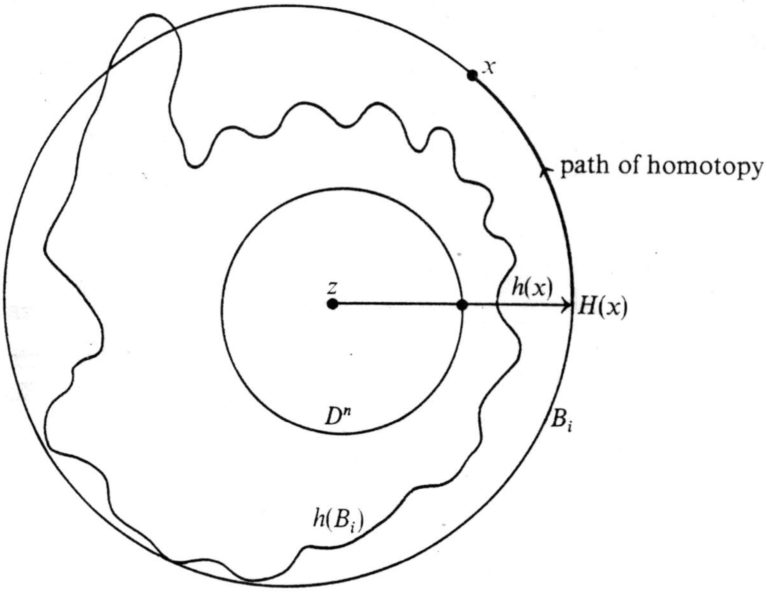

Figure 2–2.

Exercises

1. The space $C^r_S(\mathbb{R},\mathbb{R})$, $r \geqslant 0$, does not have a countable base at any point and so is not metrizable. Under the usual operations it is a topological group but not a topological vector space.

2. Let a sequence $\{f_n\}$ converge to g in $C^r_S(M,N)$, $r \geqslant 0$. Then there exists a compact set $K \subset M$ such that $f_n(x) = g(x)$ for all n and all $x \in M - K$.

3. $C^{n+1}_S(\mathbb{R},\mathbb{R})$ and $C^{n+1}_W(\mathbb{R},\mathbb{R})$ are homeomorphic to $C^n_S(\mathbb{R},\mathbb{R}) \times \mathbb{R}$ and $C^n_W(\mathbb{R},\mathbb{R}) \times \mathbb{R}$ respectively, $n \geqslant 0$.

***4.** Polynomials are dense in $C^\infty_W(\mathbb{R},\mathbb{R})$ but not in $C^0_S(\mathbb{R},\mathbb{R})$.

5. The set of closed maps is closed, but not open, in $C^\infty_S(\mathbb{R},\mathbb{R})$. (A map f is *closed* if it takes closed sets onto closed sets.)

6. The set of closed immersions is open in $C^r_S(M,N)$, $r \geqslant 1$. The set of injective immersions is not open in $C^1_S(\mathbb{R}^1,\mathbb{R}^2)$.

7. A C^1 immersion $f:M \to N$ which is injective on a closed subset $K \subset M$ is injective on a neighborhood of K. In fact f has a neighborhood $\mathscr{N} \subset C^1_S(M,N)$ and K has a neighborhood $U \subset M$ such that every $g \in \mathscr{N}$ is injective on U. If K is compact $\mathscr{N}$ can be taken in $C^1_W(M,N)$.

8. The set of proper maps is open and closed in $C^0_S(M,N)$.

9. The set of maps $f \in C^1_S(M,N)$ such that $f:M \to N$ is a covering space, is open. If M and N are compact manifolds of the same dimension, without boundaries, then every submersion $M \to N$ is a covering space.

10. If $g:\mathbb{R}^n \to \mathbb{R}^n$ is a continuous map such that

$$\liminf_{|x| \to \infty} |x - g(x)|/|x| < 2$$

then g is surjective.

11. The set of neat embeddings of M in N is open in $C^r_S(M,\partial M; N,\partial N)$, $r \geqslant 1$. But neat embeddings are not closed in $C^1_S(D^1,D^2)$.

12. A base for the strong topology on $C^r(M,N)$, $0 \leqslant r < \infty$, is obtained by taking only those sets $\mathscr{N}^r(f; \Phi,\Psi,K,\varepsilon)$ where it is further required that $K = \{K_i\}_{i \in \Lambda}$ be a covering of M by compact sets.

13. Let $\mathrm{Imm}^r_K(M,N)$ be the set of C^r maps $M \to N$ which are immersive at each point of $K \subset M$, $r \leqslant 1$. If K is compact then this set is open in $C^r_W(M,N)$.

14. Let $\mathrm{Emb}^r_K(M,N)$ be the set of C^r maps $f:M \to N$, $r \geqslant 1$, such that $f|U$ is an embedding for some open set U (depending on f), $K \subset U \subset M$. If K is compact then $\mathrm{Emb}^r_K(M,N)$ is open in $C^r_W(M,N)$.

15. Let M, N be C^r manifolds, $0 \leqslant r \leqslant \infty$. Let $\{U_i\}_{i \in \Lambda}$ be a locally finite family of open subsets of M. For each $i \in \Lambda$ let $\mathscr{A}_i \subset C^r_W(U_i,N)$ be an open set. Then the set of C^r maps $f:M \to N$ such that $f|U_i \in \mathscr{A}_i$ for all i is open in $C^r_S(M,N)$.

16. Let M, N be C^r manifolds $0 \leqslant r \leqslant \infty$. Let $V \subset M$ be an open set.
(a) The restriction map

$$\delta: C^r(M,N) \to C^r(V,N),$$
$$\delta(f) = f|V$$

is continuous for the weak topologies but not always for the strong. On the other hand:
*(b) δ is open for the strong topologies, but not always for the weak.

2. Approximations

In this section all manifolds are without boundary.

Our first job is to find a C^∞ map $\lambda:\mathbb{R}^n \to [0,1]$ with the following properties, for any given $b > a > 0$:

$$\begin{array}{llll} \text{(i)} & \lambda(x) = 1 & \text{if} & |x| \leqslant a, \\ \text{(ii)} & 1 > \lambda(x) > 0 & \text{if} & a < |x| < b, \\ \text{(iii)} & \lambda(x) = 0 & \text{if} & |x| \geqslant b. \end{array}$$

Such a map is sometimes called a *bump function*.

We start with the C^∞ map $\alpha:\mathbb{R} \to \mathbb{R}$, showing its graph on the right:

$$\alpha(x) = \begin{cases} 0 & \text{if} \quad x \leqslant 0 \\ e^{-1/x^2} & \text{if} \quad x > 0 \end{cases}$$

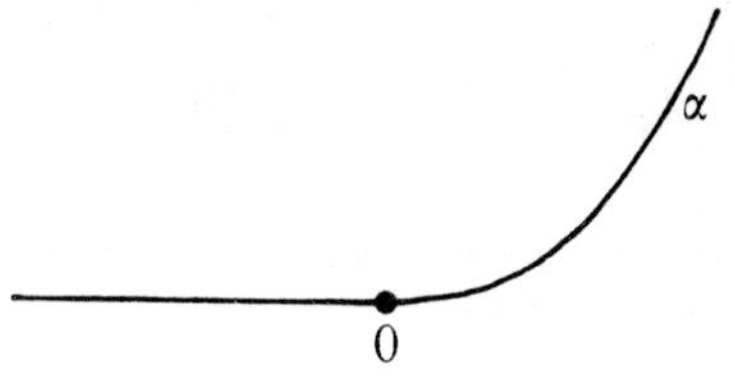

Next define $\beta:\mathbb{R} \to \mathbb{R}$:

$$\beta(x) = \alpha(x - a)\alpha(b - x)$$

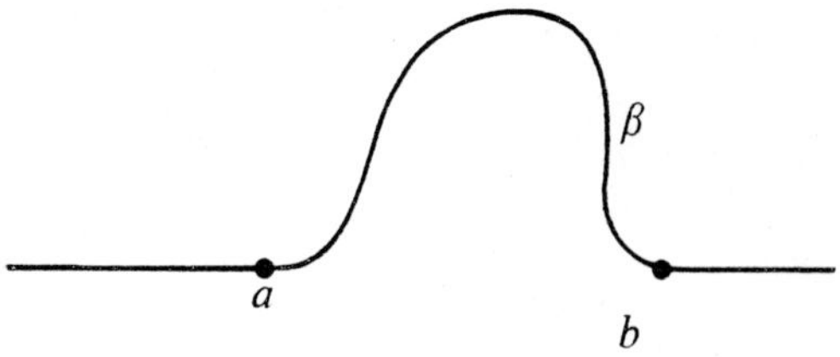

Then define $\gamma:\mathbb{R} \to [0,1]$,

$$\gamma(x) = \int_x^b \beta \Big/ \int_a^b \beta$$

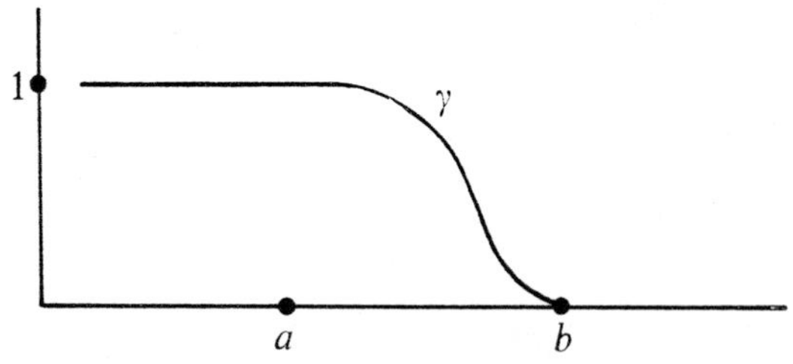

Finally define $\lambda:\mathbb{R}^n \to [0,1]$ by

$$\lambda(x) = \gamma(|x|).$$

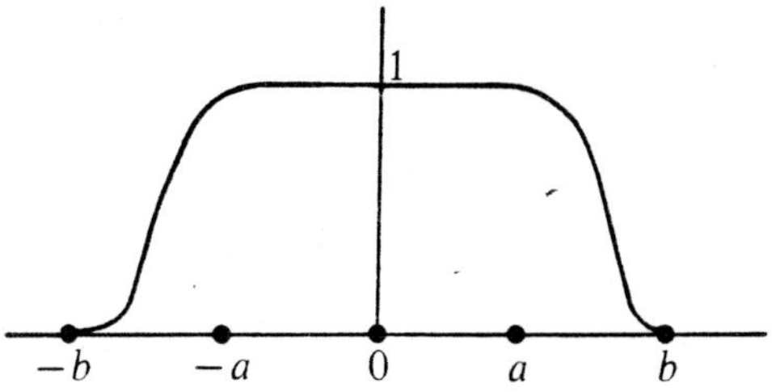

Next we define the *support* Supp f of a continuous real valued function f to be the closure of $f^{-1}(\mathbb{R} - 0)$. The complement of Supp f is the largest open set on which f vanishes.

Let M be a C^r manifold, $0 \leqslant r \leqslant \infty$, and $\mathscr{U} = \{U_i\}_{i \in \Lambda}$ an open cover. A *C^r partition of unity subordinate to* $\mathscr{U}$ is a family of C^r maps $\lambda_i : M \to [0,1]$, $i \in \Lambda$ such that the following conditions hold:

$$\text{Supp}\ \lambda_i \subset U_i \qquad (i \in \Lambda),$$

$$\{\text{Supp}\ \lambda_i\}_{i \in \Lambda} \text{ is locally finite,}$$

and

$$\textstyle\sum_{i \in \Lambda} \lambda_i(x) = 1 \qquad (x \in M).$$

Local finiteness ensures that each point has a neighborhood on which all but finitely many λ_i are 0; therefore the sum is locally a finite sum.

The third condition ensures that

$$M = \textstyle\bigcup_i \text{Int Supp}\ \lambda_i.$$

(The interior of any set S is denoted by Int S.) Therefore $\{\text{Int Supp}\ \lambda_i\}_{i \in \Lambda}$ is a locally finite open cover of M which is a shrinking of $\mathscr{U}$. (A cover $\mathscr{V} = \{V_i\}_{i \in \Lambda}$ is a *shrinking* of $\mathscr{U}$ if each $\bar{V}_i \subset U_i$.)

The following remark is often useful. *If $\mathscr{V} = \{V_\alpha\}_{\alpha \in A}$ is an open cover of M which refines $\mathscr{U} = \{U_i\}_{i \in \Lambda}$, and if $\mathscr{V}$ has a subordinate C^r partition of unity, then so has $\mathscr{U}$.* For let $\{\lambda_\alpha\}_{\alpha \in A}$ be subordinate to $\mathscr{V}$. Let $f : A \to \Lambda$ be such that $V_\alpha \in U_{f(\alpha)}$ and define

$$\mu_i : M \to [0,1],$$

$$\mu(x)_i = \textstyle\sum\{\lambda_\alpha(x);\ \alpha \in f^{-1}(i)\}.$$

Then $\{\text{Supp}\ \mu_i\}_{i \in \Lambda}$ is locally finite, for $\mu_i(x) \neq 0$ only if $i = f(\alpha)$ with $\lambda_\alpha(x) \neq 0$. Clearly Supp $\mu_i \subset U_i$ and $\sum_i \mu_i(x) = \sum_\alpha \lambda_\alpha(x) \equiv 1$.

The following theorem, one of the basic tools of differential topology, is frequently used to reduce global theorems to local form. There can be no similar theorem for analytic maps, which is why they are so much harder to handle.

2.1. Theorem. *Let M be a C^r manifold, $0 \leqslant r \leqslant \infty$. Every open cover of M has a subordinate C^r partition of unity.*

Proof. Let $\mathscr{U} = \{U_i\}_{i \in \Lambda}$ be an open cover of M. There is a locally finite atlas on M, $\{\varphi_\alpha, V_\alpha\}_{\alpha \in A}$ such that $\{\bar{V}_\alpha\}_{\alpha \in A}$ refines $\mathscr{U}$; and we may assume that each $\varphi_\alpha(V_\alpha) \subset \mathbb{R}^n$ is bounded and each $\bar{V}_\alpha \subset M$ is compact. There is a shrinking $\{W_\alpha\}_{\alpha \in A}$ of $\mathscr{V} = \{V_\alpha\}_{\alpha \in A}$, and each $\bar{W}_\alpha \subset V_\alpha$ is compact. It suffices to find a C^r partition of unity subordinate to $\mathscr{V}$.

For each $\alpha \in A$, cover the compact set $\varphi_\alpha(\bar{W}_\alpha) \subset \mathbb{R}^n$ by a finite number of closed balls

$$B(\alpha,1), \ldots, B(\alpha,k(\alpha))$$

contained in $\varphi_\alpha(V_\alpha)$. Choose C^∞ maps

$$\lambda_{\alpha,j} : \mathbb{R}^n \to [0,1], \qquad j = 1, \ldots, k(\alpha),$$

such that

$$\lambda_{\alpha,j}(x) > 0 \qquad \text{if and only if} \qquad x \in \operatorname{Int} B(\alpha,j).$$

Put

$$\lambda_\alpha = \sum_{j=1}^{k(a)} \lambda_{\alpha,j} : \mathbb{R}^n \to [0,\infty).$$

Then

$$\lambda_\alpha(x) > 0 \qquad \text{if} \qquad x \in \varphi_\alpha(\overline{W}_\alpha),$$
$$\lambda_\alpha(x) = 0 \qquad \text{if} \qquad x \in \mathbb{R}^n - \textstyle\bigcup_j B(\alpha,j).$$

Put

$$\mu_\alpha : M \to [0,\infty),$$
$$\mu_\alpha(x) = \begin{cases} \lambda_\alpha(\varphi_\alpha(x)) & \text{if} \quad x \in V_\alpha \\ 0 \quad \text{if} & x \in M - V_\alpha. \end{cases}$$

Then μ_α is C^r, $\mu_\alpha > 0$ on $\overline{W}_\alpha$, and Supp $\mu_\alpha \subset V_\alpha$. Define $v_\alpha = \mu_\alpha / \sum_\alpha \mu_\alpha$. Then $\{v_\alpha\}_{\alpha \in A}$ is a C^r partition of unity subordinate to $\mathscr{V}$.

QED

A partition of unity is used to glue together locally defined maps into $\mathbb{R}^n$ to make a globally defined map. For instance, if $\{\lambda_i\}_{i \in \Lambda}$ is a C^s partition of unity subordinate to an open cover $\{U_i\}_{i \in \Lambda}$ of M, and $g_i : U_i \to \mathbb{R}^n$ is C^s for each i, we can define

$$g : M \to \mathbb{R}^n,$$
$$g(x) = \sum \lambda_i(x) g_i(x),$$

summed over $\{i \in \Lambda : x \in U_i\}$. This is a well-defined C^s map since each x has an open neighborhood on which $\lambda_i = 0$ except for finitely many i.

The following theorem shows how the condition $\sum_i \lambda_i = 1$ can be used to obtain approximations.

2.2. Theorem. *Let M be a C^s manifold, $1 \leqslant s \leqslant \infty$. Then $C^s(M,\mathbb{R}^n)$ is dense in $C^0_S(M,\mathbb{R}^n)$.*

Proof. Let $\{V_\alpha\}_{\alpha \in A}$ be a locally finite open cover of M and for each $\alpha \in A$ let $\varepsilon_\alpha > 0$. Let $f : M \to \mathbb{R}^n$ be continuous, and suppose we want a C^s map g to satisfy $|f - g| < \varepsilon_\alpha$ on V_α for all α. For each $x \in M$ let $W_x \subset M$ be a neighborhood of x meeting only finitely many V_α. Set

$$\delta_x = \min\{\varepsilon_\alpha : x \in V_\alpha\} > 0.$$

Let $U_x \subset W_x$ be an open neighborhood of x so small that

$$|f(y) - f(x)| < \delta_x, \qquad y \in U_x.$$

Define constant maps

$$g_x : U_x \to F,$$
$$g_x(y) = f(x).$$

Relabeling the cover $\{U_x\}$ and the maps $\{g_x\}$, we have shown: there is an open cover $\{U_i\}_{i \in \Lambda} = \mathscr{U}$ of M and C^s maps $g_i: M \to F$, such that whenever $y \in U_i \cap V_\alpha$, then

$$|g_i(y) - f(y)| < \varepsilon_\alpha.$$

Let $\{\lambda_i\}_{i \in \Lambda}$ be a C^s partition of unity subordinate to $\mathscr{U}$. Define

$$g: M \to F,$$
$$g(y) = \sum_i \lambda_i(y) g_i(y).$$

Then g is C^s and

$$|g(y) - f(y)| = \left|\sum \lambda_i(y) g_i(y) - \sum \lambda_i(y) f(y)\right| \leqslant \sum \lambda_i(y) |g_i(y) - f(y)|.$$

Hence if $y \in V_\alpha$,

$$|g(y) - f(y)| < \sum \lambda_i(y) \varepsilon_\alpha = \varepsilon_\alpha.$$

QED

Our next task is to approximate C^r maps by C^s maps, $s > r \geqslant 1$, in the strong C^r topology. The preceding argument will not work now because the derivatives of the λ_i are involved. We need to uniformly approximate $f|U_i$ not by constants, but by C^r maps whose derivatives up to order r uniformly approximate those of f. For maps defined on open subsets of $\mathbb{R}^n$ we achieve this by the technique of convolution, discussed next.

Let $\theta: \mathbb{R}^m \to \mathbb{R}$ be a map having compact support. There is a smallest $\sigma \geqslant 0$ such that Supp θ is contained in the closed ball $B_\sigma(0) \subset \mathbb{R}^m$ of radius σ and center 0. We call σ the *support radius* of θ.

Let $U \subset \mathbb{R}^m$ be open and $f: U \to \mathbb{R}^n$ a map. If $\theta: \mathbb{R}^m \to \mathbb{R}$ has compact support we define the *convolution of f by θ* to be the map

$$\theta * f: U_\sigma \to \mathbb{R}^n$$

given by

(1) $$\theta * f(x) = \int_{B_\sigma(0)} \theta(y) f(x - y)\, dy \qquad (x \in U_\sigma)$$

where

$$U_\sigma = \{x \in U: B_\sigma(x) \subset U\}.$$

The integral is the Lebesgue integral, dy denoting the usual measure on $\mathbb{R}^m$.

The integrand in (1) is 0 on the boundary of $B_\sigma(0)$; we extend it to a continuous map $\mathbb{R}^m \to \mathbb{R}$ by defining it to be 0 outside $B_\sigma(0)$. Therefore we have

(2) $$\theta * f(x) = \int_{\mathbb{R}^m} \theta(y) f(x - y)\, dy \qquad (x \in U_\sigma).$$

For a fixed $x \in U_\sigma$ we make the measure preserving change of variable in (1): $z = x - y$. Then

(3) $$\theta * f(x) = \int_{B_\sigma(x)} \theta(x - z) f(z)\, dz$$
$$= \int_{\mathbb{R}^m} \theta(x - z) f(z)\, dz \qquad (x \in U_\sigma)$$

where again the integrand is defined to be 0 outside $B_\sigma(x)$.

A map $\theta:\mathbb{R}^m \to \mathbb{R}$ is called a *convolution kernel* if it is nonnegative, has compact support, and $\int_{\mathbb{R}^m} \theta_m = 1$. It is clear that there exist C^∞ convolution kernels of any given support radius.

We may think of $\theta * f(x)$ as a weighted average of the values of f near x. This makes it plausible that $\theta * f$ will be an approximation to f which will be smooth if θ is smooth.

We introduce the notation

$$\|f\|_{r,K} = \sup\{\|D^k f(x)\|: x \in K, 0 \leqslant k \leqslant r\}$$

if $f:U \to \mathbb{R}^n$ is C^r, $U \subset \mathbb{R}^m$ is open, $K \subset U$ is any subset, and $\|D^k f(x)\|$ is the norm of the k'th derivative of f at x. Here $\|D^0 f(x)\|$ means $|f(x)|$, while for $k > 0$ the k'th derivative at x is a k-linear map from $\mathbb{R}^m$ to $\mathbb{R}^n$. The norm $\|S\|$ of any k-linear map

$$S:\mathbb{R}^m \times \cdots \times \mathbb{R}^m \to \mathbb{R}^n$$

is defined to be the maximum of $|S(x_1, \ldots, x_k)|$ where the *vectors* $x_1, \ldots, x_k$ vary over the unit sphere in $\mathbb{R}^m$. The value ∞ is allowed for $\|f\|_{r,K}$. If K is the entire domain of f we write simply $\|f\|_r$. Note that for all $y_1, \ldots, y_k$ in $\mathbb{R}^m$:

$$|S(y_1, \ldots, y_k)| \leqslant \|S\| \cdot |y_1| \cdots |y_k|.$$

We have the following basic result:

2.3. Theorem. *Let $\theta:\mathbb{R}^m \to \mathbb{R}$ have support radius $\sigma > 0$. Let $U \subset \mathbb{R}^m$ be an open set, and $f:U \to \mathbb{R}^n$ a continuous map. The convolution $\theta * f: U_\sigma \to F$ has the following properties*:

(a) *If $\theta|$Int Supp θ is C^k, $1 \leqslant k \leqslant \infty$, then so is $\theta * f$; and for each finite k,*

$$D^k(\theta * f) = (D^k\theta) * fx$$

on U_σ.

(b) *If f is C^k then*

$$D^k(\theta * f) = \theta * (D^k f).$$

(c) *Suppose f is C^r, $0 \leqslant r \leqslant \infty$. Let $K \subset U$ be compact. Given $\varepsilon > 0$ there exists $\sigma > 0$ such that $K \subset U_\sigma$, and if θ is a C^r convolution kernel of support radius σ, then $\theta * f$ is C^r and*

$$\|\theta * f - f\|_{r,K} < \varepsilon.$$

Proof. To prove (b), observe that the domain of integration in (1) can be restricted to Int Supp θ; the integrand is then differentiable in x, and (b) follows by induction on k and differentiating under the integral sign. (a) is proved similarly using (3). To prove (c) it suffices to take $r = 0$, by (b). Since $d(K,\mathbb{R}^m - U) > 0$, we can choose σ so small that $K \subset U_\sigma$. By uniform continuity of f on a compact neighborhood of K, we can choose σ so small that if $x \in K$ and $|x - y| \leqslant \sigma$ then $|f(x) - f(y)| < \varepsilon$.

Since

$$\int_{\mathbb{R}^m} \theta = 1$$

we have, integrating over $\mathbb{R}^m$:

$$
\begin{aligned}
|\theta * f(x) - f(x)| &= \left|\int \theta(y)(f(x-y) - f(x))\,dy\right| \\
&\leqslant \int \theta(y)|f(x-y) - f(x)|\,dy \\
&\leqslant \varepsilon \int \theta(y)\,dy = \varepsilon.
\end{aligned}
$$

QED

From Theorem 2.3 it immediately follows that a C^r map from open subset of $\mathbb{R}^m$ to $\mathbb{R}^n$ can be C^r approximated by C^∞ maps in neighborhoods of compact sets. Using partitions of unity we prove the following stronger approximation theorem:

2.4. Theorem. *Let $U \subset \mathbb{R}^m$ and $V \subset \mathbb{R}^n$ be open sets. Then $C^\infty(U,V)$ is dense in $C_S^r(U,V)$, $0 \leqslant r < \infty$.*

Proof. Since $C_S^r(U,V)$ is open in $C_S^r(U,\mathbb{R}^n)$, it suffices to prove the theorem with $V = \mathbb{R}^n$.

Let $f \in C^r(U,\mathbb{R}^n)$. A neighborhood base at f in $C_S^r(U,\mathbb{R}^n)$ consists of sets $\mathcal{N}(f,K,\varepsilon)$ of the following form (see Exercise 12, Section 2.1). Let $K = \{K_i\}_{i \in \Lambda}$ be a locally finite family of compact sets covering U; let $\varepsilon = \{\varepsilon_i\}_{i \in \Lambda}$ be a family of positive numbers, and let $\mathcal{N}(f,K,\varepsilon)$ be the set of C^r maps $g: U \to \mathbb{R}^n$ such that for all $i \in \Lambda$

(4) $$\|g - f\|_{r, K_i} < \varepsilon_i.$$

Fixing f,K and ε, we must show

$$C^\infty(U,\mathbb{R}^n) \cap \mathcal{N}(f,K,\varepsilon) \neq \varnothing.$$

Let $\{\lambda_i\}_{i \in \Lambda}$ be a C^∞ partition of unity on U such that Supp λ_i is compact and contains K_i.

Given positive numbers $\{\alpha_i\}_{i \in \Lambda}$ there are C^∞ maps $g_i: U_i \to \mathbb{R}$ such that for $x \in K_i$ and $k = 0, \ldots, r$,

$$\|g_i - f\|_{r, K_i} < \alpha_i.$$

Put

$$
\begin{gathered}
g: U \to F \\
g(x) = \textstyle\sum_i \lambda_i(x) g_i(x).
\end{gathered}
$$

Then g is C^∞. To estimate $\|D^k g(x) - D^k f(x)\|$ we observe that if $\lambda: U \to R$ and $\varphi: U \to F$ are C^k and $\psi(x) = \lambda(x)\varphi(x)$, then $D^k\psi(x)$ is a bilinear function of $D^p\lambda(x)$, $D^q\varphi(x)$, p, $q = 0, \ldots, k$; *this bilinear function is universal and independent of x, λ and φ*. Thus there is a universal constant $A_k > 0$ such that

$$\|D^k(\lambda\varphi)(x)\| \leqslant A_k \max_{0 \leqslant p \leqslant k} \|D^p\lambda(x)\| \cdot \max_{0 \leqslant q \leqslant k} \|D^q\varphi(x)\|.$$

Set

$$A = \max\{A_0, \ldots, A_r\}.$$

Fix $i \in \Lambda$ and set

$$\Lambda_i = \{j \in \Lambda : K_i \cap K_j \neq \varnothing\}.$$

This is a finite set; let its cardinality be m_i. Put

$$\mu_i = \max \{\|\lambda_j\|_{r, K_i} : j \in \Lambda_i\}$$
$$\beta_i = \max \{\alpha_j : j \in \Lambda_i\}.$$

In the following sums j varies over Λ_i. For $x \in K_i$ and $0 \leqslant k \leqslant r$ we have

$$\|D^k g(x) - D^k f(x)\| = \|\textstyle\sum_j D^k(\lambda_j g_j - \lambda_j f)(x)\|$$
$$\leqslant \textstyle\sum_j \|D^k(\lambda_j(g_j - f))(x)\| \leqslant m_i A \mu_i \beta_i.$$

It is clear that the numbers α_i can be chosen so that

$$m_i A \mu_i \beta_i < \varepsilon_i.$$

With this choice of α_i we have, for all $i \in \Lambda$,

$$\|g - f\|_{r, K_i} < \varepsilon_i.$$

QED

A refinement of Theorem 2.4 is needed for the globalization to manifolds. Suppose we have a C^r map $f: U \to \mathbb{R}^n$ that we want to approximate by a C^r map h which is C^s in a neighborhood of a (relatively) closed set $K \subset U$. At the same time, for technical reasons we want h to equal f outside a certain open set $W \subset U$; of course, we must assume f is already C^s on a neighborhood of $K - W$. The following *relative approximation* theorem ensures that maps such as h approximate f arbitrarily closely.

2.5. Theorem. *Let $U \subset \mathbb{R}^m$ and $V \subset \mathbb{R}^n$ be open subsets, and $f: U \to V$ a C^r map. Let $K \subset U$ be closed and $W \subset U$ open, such that f is C^s on a neighborhood of $K - W$. Then every neighborhood $\mathcal{N}$ of f in $C^r_S(U,V)$ contains a C^r map $h: U \to V$ which is C^s in a neighborhood of K, and which equals f on $U - W$.*

Proof. We may assume $V = \mathbb{R}^n$ (see proof of Theorem 2.4). Let $A \subset U$ be an open set containing the closed set $K - W$ such that $f|A$ is C^s. Let $W_0 \subset U$ be open, with

$$K - A \subset W_0 \subset \overline{W}_0 \subset W.$$

Let $\{\lambda_0, \lambda_1\}$ be a C^s partition of unity for the open cover $\{W, U - \overline{W}_0\}$ of U. Thus λ_0 and λ_1 are C^s maps $U \to [0,1]$ such that $\lambda_0 + \lambda_1 = 1$, $\lambda_0 = 0$ on a neighborhood of $U - W$ and $\lambda_1 = 0$ on a neighborhood of $\overline{W}_0$.

Define

$$G: C^r_S(U,\mathbb{R}^n) \to C^r_S(U,\mathbb{R}^n)$$

by

$$G(g)(x) = \lambda_0(x)g(x) + \lambda_1(x)f(x).$$

Then $G(g) = g$ in W_0 and $G(g) = f$ in $U - W$. Clearly $G(g)$ is C^s on every open set on which both f and g are C^s. It is easy to prove G continuous. Since $G(f) = f$, there is an open set $\mathcal{N}_0 \subset C^r_S(U,\mathbb{R}^n)$ containing f such that

$G(\mathcal{N}_0) \subset \mathcal{N}$. By Theorem 2.4 there is a C^s map $g \in \mathcal{N}_0$. Then $h = G(g)$ has the required properties.

QED

We now prove the basic approximation theorem for manifolds without boundary. Later we shall extend it to ∂-manifolds and manifold pairs.

2.6. Theorem. *Let M and N be C^s manifolds, $1 \leqslant s \leqslant \infty$. Then $C^s(M,N)$ is dense in $C^r_S(M,N)$, $0 \leqslant r < s$.*

Proof. Let $f: M \to N$ be C^r. Let $\Phi = \{\varphi_i, U_i\}_{i \in \Lambda}$ be a locally finite atlas for M and $\Psi = \{\psi_i, V_i\}_{i \in \Lambda}$ a family of charts for N such that for all $i \in \Lambda$, $f(U_i) \subset V_i$. Let $L = \{L_i\}_{i \in \Lambda}$ be a closed cover of M, $L_i \subset U_i$. Let $\varepsilon = \{\varepsilon_i\}_{i \in \Lambda}$ be a family of positive numbers, and put $\mathcal{N} = \mathcal{N}^r(f; \Phi, \Psi, L, \varepsilon) \subset C^r(M,N)$.

We look for a $g \in \mathcal{N}$ which is C^s. The set Λ is countable; we therefore assume that $\Lambda = \mathbb{Z}_+$ or, if M is compact, $\Lambda = \{1, \ldots, p\}$. (We denote the integers by $\mathbb{Z}$, the positive integers by $\mathbb{Z}_+$, and the natural numbers by $\mathbb{N} = \mathbb{Z}_+ \cup \{0\}$.)

Let $\{W_i\}_{i \in \Lambda}$ be a family of open sets in M such that $L_i \subset W_i \subset \overline{W}_i \subset U_i$.

We shall define by induction a family of C^r maps $g_k \in \mathcal{N}$, $k \in \mathbb{N}$, having the following properties: $g_0 = f$, and for $k \geqslant 1$:

$$(5)_k \qquad g_k = g_{k-1} \text{ on } M - W_k$$

$$(6)_k \qquad g_k \text{ is } C^s \text{ in a neighborhood of } \textstyle\bigcup_{0 \leqslant j \leqslant k} L_j.$$

Assuming for the moment that the g_k exist, define $g: M \to N$ by $g(x) = g_{\kappa(x)}(x)$, where $\kappa(x) = \max\{k : x \in \overline{U}_k\}$. Each x has a neighborhood on which $g = g_{\kappa(x)}$. This shows that g is C^s and $g \in \mathcal{N}$, and the theorem is proved.

It remains to construct the g_k. Put $g_0 = f$; then $(5)_0$ and $(6)_0$ are vacuously true. Suppose that $0 < m$ and that we have maps $g_k \in \mathcal{N}$, $0 \leqslant k < m$, satisfying $(5)_k$ and $(6)_k$.

Define a space of maps

$$\mathcal{G} = \{h \in C^r_S(U_m, V_m) : h = g_{m-1} \text{ on } U_m - W_m\}.$$

Define

$$T: \mathcal{G} \to C^r_S(M,N),$$

$$T(h) = \begin{cases} h & \text{on} \quad U_m \\ g_{m-1} & \text{on} \quad M - U_m. \end{cases}$$

Then T is easily proved continuous. Observe that $T(g_{m-1}|U_m) = g_{m-1}$. Hence $T^{-1}(\mathcal{N}) \neq \varnothing$.

Let $K = \bigcup_{k \leqslant m} L_k \cap U_m$. Then K is a closed subset of U_m and $g_{m-1}: U_m \to V_m$ is C^s on a neighborhood of $K - W_m$. Since U_m and V_m are C^r diffeomorphic to open subsets of $\mathbb{R}^m$ and $\mathbb{R}^n$, we can apply Theorem 2.5 to $C^r_S(U_m, V_m)$. We conclude that the maps in $\mathcal{G}$ which are C^s in a neighborhood of K are dense in $\mathcal{G}$. Therefore $T^{-1}(\mathcal{N})$ contains such a map h. Define $g_m = T(h)$; then $g_m \in \mathcal{N}$ satisfies $(5)_m$ and $(6)_m$.

QED

For the following application of Theorem 2.6 let $G^k(M,N) \subset C^k(M,N)$, $k \geqslant 1$, denote any one of the following subsets:

diffeomorphisms,
embeddings,
closed embeddings,
immersions,
submersions,
proper maps.

2.7. Theorem. *Let M and N be C^s manifolds, $1 \leqslant s \leqslant \infty$, with* $\dim M < \infty$. *Then $G^s(M,N)$ is dense in $G^r(M,N)$ in the strong C^r topology, $1 \leqslant r < s$. In particular, M and N are C^s diffeomorphic if and only if they are C^r diffeomorphic.*

Proof. This follows from Theorem 2.6 and the openness theorems of Section 2.1.

QED

We shall need the following lemma for raising differentiability of manifolds:

2.8. Lemma. *Let U be a C^r manifold, $0 \leqslant r < \infty$, and $W \subset U$ an open set. Let $V \subset \mathbb{R}^n$ be open, $f \in C^r_S(U,V)$, and put $f(W) = V'$. Then there is a neighborhood $\mathcal{N} \subset C^r(W,V')$ of $f|W$ such that if $g_0 \in \mathcal{N}$, the map*

$$T(g_0) = g: U \to V,$$
$$g = \begin{cases} g_0 & \text{on} \quad W \\ f & \text{on} \quad U - W \end{cases}$$

is C^r, and $T: \mathcal{N} \to C^r_S(U,V)$ is continuous.

Proof. Let $\{\varphi_i, U_i\}_{i \in \Lambda}$ be a locally finite family of charts of U which cover the boundary Bd W of W. Let $\{L_i\}_{i \in \Lambda}$ be a family of closed subsets of U and which cover Bd W, with $L_i \subset U_i$.

Let $\mathcal{N} \subset C^r_S(W,V')$ be the set of C^r maps $h: W \to V'$ such that if $i \in \Lambda$, $y \in \varphi_i(L_i)$ and $0 \leqslant k \leqslant r$, then

$$\|D^k(h\varphi_i^{-1})y - D^k(f\varphi_i^{-1})y\| < d(y, \varphi_i(U_i - W)). \tag{7}$$

Then $\mathcal{N}$ is a neighborhood of $f|W$: by paracompactness W has a locally finite closed cover $\{K_\alpha\}$ such that each K_α meets only finitely many L_i and on each $K_\alpha \cap L_i$ the map $x \mapsto d(\varphi_i(x), \varphi_i(U_i - W))$ is bounded away from 0.

If $h \in \mathcal{N}$ define

$$T(h) = g: U \to V,$$
$$g = \begin{cases} h & \text{on} \quad W \\ f & \text{on} \quad U - W. \end{cases}$$

We claim g is C^r. It suffices to prove that each map $\lambda_i = (g - f)\varphi_i^{-1}: \varphi_i(U_i) \to \mathbb{R}^n$ is C^r. Now

$$\lambda_i = \begin{cases} h\varphi_i^{-1} - f\varphi_i^{-1} & \text{on} \quad \varphi_i(W) \\ 0 & \text{on} \quad \varphi_i(U_i - W). \end{cases}$$

Obviously λ_i is C^r on $\varphi_i(W)$. By (7), for $0 \leqslant k \leqslant r$:

$$D^k\lambda_i(y) \mapsto 0 \qquad \text{as} \qquad d(y,\varphi_i(U_i - W)) \to 0,$$

uniformly in $y \in \varphi_i(W)$. It follows that λ_i is C^r, with all derivatives 0 on $\varphi_i(U_i - W)$. Therefore g is C^r and the map $T: C^r(W,V') \to C^r(U,V)$ is well-defined. The continuity of T is left as an exercise.

QED

Let α be a C^r differential structure on a manifold M. A C^s differential structure β on M, $s > r$, is *compatible with* α if $\beta \subset \alpha$. This means that every chart of M_β is a chart of M_α. Equivalently, it means that the identity map of M is a C^r diffeomorphism $M_\alpha \to M_\beta$.

2.9. Theorem. *Let α be a C^r differential structure on a manifold M, $r \geqslant 1$. For every s, $r < s \leqslant \infty$, there exists a compatible C^s differential structure $\beta \subset \alpha$, and β is unique up to C^s diffeomorphism.*

Proof. For convenience we shall denote a differential structure and its restriction to an open set by the same symbol. By Zorn's lemma there is a nonempty open set $B \subset M$ and a C^s differential structure β on B which is compatible with α, and such that (B,β) is maximal in this property. We must prove $B = M$.

If $B \neq M$ there is a chart (φ,U) for M such that $U \cap (M - B) \neq \varnothing$. Put $\varphi(U) = U' \subset \mathbb{R}^n$, $U \cap B = W$ and $\varphi(W) = W'$.

There are now two differential structures on W: the C^r structure α and the compatible C^s structure $\beta \subset \alpha$. We shall find a C^r diffeomorphism $\theta: U_\alpha \to U'$ such that $\theta|W: W_\beta \to W'$ is a C^s diffeomorphism. In that case the chart (θ,U) has C^s overlap with β; the C^s atlas $\beta \cup (\theta,U)$ on $B \cup U$ is contained in α and this contradicts the maximality of (B,β).

To construct θ we use Theorem 2.8 to obtain a neighborhood $\mathcal{N} \subset C^r_S(W_\beta,W')$ of $\varphi|W: W \to W'$ with the following property. Whenever $\psi_0 \in \mathcal{N}$, the map $T(\psi_0) = \psi: U \to U'$ defined by

$$\psi = \begin{cases} \psi_0 & \text{on} \quad W \\ \varphi & \text{on} \quad U - W \end{cases}$$

is C^r, and the resulting map

$$T: C^r_S(W_\beta,W') \to C^r_S(U,U')$$

is continuous. Since $T(\varphi|W)$ is the diffeomorphism φ, there is a neighborhood $\mathcal{N}_0 \subset \mathcal{N}$ of $\varphi|W$ such that $T(\mathcal{N}_0) \subset \mathrm{Diff}^r(U,U')$. By the approximation theorem 2.6 there is a C^s diffeomorphism $\theta_0 \in \mathcal{N}_0$; the required map θ is then $T(\theta_0)$.

QED

It is amusing that neither paracompactness nor Hausdorffness of M was used in the preceding proof; these properties were needed only for the coordinate domain U.

From Theorems 2.7 and 2.9 we obtain a fundamental result:

2.10. Theorem.

(a) *Let* $1 \leqslant r < \infty$. *Every* C^r *manifold is* C^r *diffeomorphic to a* C^∞ *manifold.*

(b) *Let* $1 \leqslant r < s \leqslant \infty$. *If two* C^s *manifolds are* C^r *diffeomorphic, they are* C^s *diffeomorphic.*

In view of this there is no need to consider C^r manifolds for $1 \leqslant r < \infty$; and for most purposes C^∞ maps are sufficient. The C^∞ category has several advantages over the C^r categories with r finite. An obvious one is its closure under derivatives. A more subtle advantage comes from the Morse–Sard theorem in the following chapter.

Earlier we pointed out the theme of *globalization* that runs through many proofs. There is an abstract aspect of the passage from local to global which, if recognized, can often make a proof obvious; at the least, it provides a clear strategy for the proof. Since we shall need to globalize several more times, it is worthwhile to formalize the pattern of proof in the following way.

Let X be a set and $\mathfrak{Y}$ a family of subsets. We assume that $X \in \mathfrak{Y}$ and that the union of any collection of elements of $\mathfrak{Y}$ is again in $\mathfrak{Y}$. (In practice X is a manifold and $\mathfrak{Y}$ is generated by the elements in an open cover, or a locally finite closed cover.) Suppose we have a contravariant functor from the partially ordered (by inclusion) set $\mathfrak{Y}$ to the category of sets. That is, to every $A \in \mathfrak{Y}$ there is associated a set $\mathscr{F}_A$, and to every pair of sets $A, B \in \mathfrak{Y}$ with $A \subset B$ there is assigned a map of sets $\mathscr{F}_{AB}: \mathscr{F}(B) \to \mathscr{F}(A)$, denoted by $x \mapsto x|A$, where $x \in \mathscr{F}(B)$ such that $\mathscr{F}_{AB}\mathscr{F}_{BC} = \mathscr{F}_{AC}$ whenever $A \subset B \subset C$, and $\mathscr{F}_{AA} =$ identity map of $\mathscr{F}(A)$.

We call $(\mathscr{F}, \mathfrak{Y})$ a *structure functor* on X. An element of $\mathscr{F}(A)$ is thought of as a "structure" of some kind on A for which "restriction" to subsets of A makes sense. We wish to prove that X has a structure, that is, $\mathscr{F}(X) \neq \varnothing$.

A structure functor is *continuous* if the following holds. If $\{Y_\alpha\}$ is any simply ordered family of elements of $\mathfrak{Y}$, and $\cup Y_\alpha = Y$, then the inverse limit of the maps $\mathscr{F}_{Y_\alpha Y}: \mathscr{F}(Y) \to \mathscr{F}(Y_\alpha)$ is a map $\mathscr{F}(Y) \to \operatorname{inv\,lim} \mathscr{F}(Y_\alpha)$ which is bijective.

A structure functor is *locally extendable* if every point of X belongs to a set $V \in \mathfrak{Y}$ such that for all $Y \in \mathfrak{Y}$ the map

$$\mathscr{F}_{V,Y \cup V}: \mathscr{F}(Y \cup V) \to \mathscr{F}(Y)$$

is surjective. It is called *nontrivial* if $\mathscr{F}(Y) \neq \varnothing$ for some $Y \in \mathfrak{Y}$.

As an example, let X be a C^r manifold, $0 \leqslant r < \infty$, and $\mathfrak{Y}$ the family of all open sets. Let $\mathscr{F}(Y)$ be the set of compatible C^s differential structures on Y for some $s > r$. Clearly $(\mathscr{F},\mathfrak{Y})$ is a structure functor which is nontrivial and continuous. If $1 \leqslant r < s \leqslant \infty$ then $\mathscr{F}$ is locally extendable. as was shown in the proof of Theorem 2.9.

For another example, let $X \subset \mathbb{R}^n$ be a nonnull open set. Let $\mathfrak{Y}$ be generated by a locally finite cover of X by compact sets. Let $\mathscr{F}(Y)$ be the set of Y-germs of analytic maps into $\mathbb{R}$; a *Y-germ* is an equivalence class of maps defined in neighborhoods of Y, two maps being equivalent if they agree in some neighborhood of Y. Then $(\mathscr{F},\mathfrak{Y})$ is continuous and nontrivial, but is not locally extendable.

2.11. Globalization Theorem. *Let $(\mathscr{F},\mathfrak{Y})$ be a nontrivial structure functor on X which is continuous and locally extendable. Then $\mathscr{F}(X) \neq \varnothing$. In fact if $\mathscr{F}(Y_0) \neq \varnothing$ then $\mathscr{F}(X) \to \mathscr{F}(Y_0)$ is surjective.*

Proof. Let $a_0 \in \mathscr{F}(Y_0)$. Let S be the set of pairs (Y,a) with $Y_0 \subset Y \in \mathfrak{Y}$ and $a \in \mathscr{F}(Y)$, $a|Y_0 = a_0$. Partially order S by $(Y',a') \leqslant (Y,a)$ if $Y' \subset Y$ and $a|Y' = a'$. By the closure of $\mathfrak{Y}$ under union and the continuity axiom, S has a maximal element $(Y*,a*)$.

We claim $Y* = X$. If not, by local extendability there exists $V \in \mathfrak{Y}$ and $b \in \mathscr{F}(Y* \cup V)$ such that $b|Y* = a*$. But then $(Y* \cup V,b) > (Y*,a*)$, contradicting maximality. Hence $Y* = X$.

QED

This method is often convenient for proving the existence of a global structure when local existence and extendability are known; Theorem 2.9 is an example. Another example is:

2.12. Theorem. *Let M, N be C^r manifolds, $1 \leqslant r \leqslant \infty$. If* $\dim N \geqslant 2 \dim M$ *then immersions are dense in $C^r_S(M,N)$.*

Proof. It is convenient to assume $r = 2$. This is no restriction, for if $r < 2$, every C^1 manifold has a compatible C^2 structure, and if $r > 2$ every C^2 immersion can be approximated by C^r immersions. Henceforth we assume all maps are C^2.

Let $f_0: M \to N$ be a map and $\mathcal{N}_0 \subset C^2_S(M,N)$ a neighborhood of f_0. There is a smaller neighborhood $\mathcal{N} \subset \mathcal{N}_0$ of f_0 of the following form:

$$\mathcal{N} = \mathcal{N}^2(f_0; \Phi,\Psi,K,\varepsilon)$$

where

$$\Psi = \{\psi_i: V_i \to \mathbb{R}^n\}_{i \in \Lambda}$$

is a family of charts on N;

$$\Phi = \{\varphi_i: U_i \to \mathbb{R}^m\}_{i \in \Lambda}$$

is a locally finite atlas for M with $f_0(U_i) \subset V_i$; $K = \{K_i\}_{i \in \Lambda}$, where $K_i \subset U_i$;

$$\varphi_i(K_i) = D^m \subset \mathbb{R}^m;$$

and $\varepsilon = \{\varepsilon_i\}_{i \in \Lambda}$ is a set of positive numbers.

We now define a structure functor $(\mathscr{F},\mathfrak{Y})$. For $\mathfrak{Y}$ we take the family of all unions of the disks K_i, partially ordered by inclusion. If $A \in \mathfrak{Y}$ we define $\mathscr{F}(A)$ to be the set of A-germs of maps $f \in \mathscr{N}$ such that f is an immersion of some open set containing A. If $A \subset B$, every such B-germ is contained in a unique A-germ; this correspondence defines the map $\mathscr{F}_{AB}:\mathscr{F}(B) \to \mathscr{F}(A)$.

It is easy to see that $(\mathscr{F},\mathfrak{Y})$ is continuous. Moreover, it is locally extendable. To see this let $W \subset M$ be an open neighborhood of $A \in \mathfrak{Y}$, and let $f \in \mathscr{N}$ be a map such that $f|W$ is an immersion. Then f represents an A-germ $\alpha \in \mathscr{F}(A)$. If $A \neq M$ pick $K_i \not\subset A$. To extend α over $A \cup K_i$ is to find a map $g \in \mathscr{N}$ which agrees with f in some neighborhood of A, and which is an immersion of a neighborhood of $A \cup K_i$. This is done as follows.

Let $B \subset U_i$ be a disk whose interior contains K_i. Now $f(B) \subset V_i$, and V_i is diffeomorphic to an open set in $\mathbb{R}^n$, $n \geqslant 2 \dim M$. Therefore by Theorem 1.1 $f|B:B \to V_i$ can be C^2 approximated by immersions.

Let $\lambda:M \to [0,1]$ equal 0 on an open neighborhood Z of $A \cup (M - \operatorname{int} B)$, and 1 on an open neighborhood Y of $K_i - W$. For each immersion $g:B \to V_i$ define a map $S(g):M \to N$ by

$$S(g)x = \begin{cases} f(x) & \text{if} \quad x \in M - B \\ (1 - \lambda(x))f(x) + \lambda(x)g(x) & \text{if} \quad x \in B. \end{cases}$$

Here we have identified V_i with an open set in $\mathbb{R}^n$ via ψ_i. The map

$$S:C_S^2(B,V_i) \to C_S^2(M,N)$$

is continuous, and

$$S(f|B) = f,$$
$$S(g) = f \quad \text{on} \quad Z$$
$$S(g) = g \quad \text{on} \quad Y.$$

From the first equation it follows that $S(g) \in \mathscr{N}$ if g is sufficiently near $f|B$. The second equation implies that $S(g)$ and f have the same A-germ. The third equation implies that $S(g)$ is an immersion on Y. But in fact $S(g)$ is an immersion of a neighborhood of K_i if g is sufficiently near $f|B$. For let X be a neighborhood of $K_i - Y$ with $\bar{X}$ compact in W. Since $f|W$ is an immersion and $\bar{X} \subset W$ is compact, f has a neighborhood $\mathscr{N}_1$ such that $h|X$ is an immersion if $h \in \mathscr{N}_1$. If g is close enough to f then $S(g) \in \mathscr{N}_1$ and so $S(g)|X$ is an immersion. Such an $S(g)$ is thus an immersion of a neighborhood of $A \cup K_i$. This proves that $(\mathscr{F},\mathfrak{Y})$ is locally extendable. By Theorem 2.11, $\mathscr{F}(M) \neq \varnothing$, that is, $\mathscr{N}$ contains an immersion.

Another proof of Theorem 2.12 is given at the end of Section 3.2.

QED

If M is not compact then embeddings may not be dense in $C_S^r(M,N)$, no matter what the dimensions of M, N. For example let $f:\mathbb{Z} \to \mathbb{R}^n$ be a map of the integers whose image is the set of points having rational coordinates.

Let $g:\mathbb{Z} \to \mathbb{R}^n$ be any map such that

$$|g(n) - f(n)| < \frac{1}{|n|}$$

for $n \neq 0$. Then the image of g is dense, so $g(\mathbb{Z})$ is not discrete; hence g is not an embedding.

The difficulty is trying to imitate the proof of Theorem 2.12 for embeddings is that if M is not compact the structure functor defined by A-germs of embeddings is not continuous. It is continuous, however, when M is compact; and if $\dim N \geqslant 2 \dim M + 1$ it can be proved locally extendible. More generally, if the neighborhood $\mathcal{N} \subset C_S^r(M,N)$ consists of proper maps, then the structure functor of A-germs of maps in $\mathcal{N}$ which are embeddings in a neighborhood of A is continuous, and locally extendable if $\dim N \geqslant 2 \dim M + 1$. In this way one obtains a proof of the following result; the details are left as an exercise:

2.13. Theorem. *Let M, N be C^r manifolds, $1 \leqslant r \leqslant \infty$, with* $\dim N \geqslant 2 \dim M + 1$. *If M is compact then embeddings are dense in $C_S^r(M,N)$. If M is not compact, embeddings are dense in* $\mathrm{Prop}_S^r(M,N)$.

As a corollary we have the following result of Whitney:

2.14. Theorem. *Every C^r n-dimensional manifold, $1 \leqslant r \leqslant \infty$, is C^r diffeomorphic to a closed submanifold of $\mathbb{R}^{2n+1}$.*

Proof. This follows from Theorem 2.13 as soon as a proper map $f:M \to \mathbb{R}$ has been found. If M is compact take f constant. If M is not compact, let $M_1 \subset M_2 \subset \cdots$ be an increasing family of compact sets which fill up M, with $M_k \subset \mathrm{Int}\, M_{k+1}$.

Let

$$f(M_{2k+1} - M_{2k}) = 2k + 1, \qquad k \geqslant 1$$

and extend f to a continuous map sending $M_{2k} - M_{2k-1}$ into $[2k - 1, 2k + 1]$ using Tietze's theorem. Then f is proper.

QED

We return to Theorem 2.13 in Section 2.4.

Exercises

*__1.__ (a) Let $X \subset M$ be a closed subset of a C^r manifold, $0 \leqslant r \leqslant \infty$. Then there exists a C^r map $f:M \to [0,\infty)$ with $X = f^{-1}(0)$. [Hint: let $f|M - X$ be a C^r map which closely approximates, in the strong C^0 topology, the map $\exp(-d(x,X))$.]

(b) If X, Y are disjoint closed subsets of a C^r manifold M, $0 \leqslant r \leqslant \infty$, there is a C^r map $\lambda:M \to [0,1]$ with $\lambda^{-1}(0) = X$, $\lambda^{-1}(1) = Y$.

2. Any two points in a connected C^r manifold, $0 \leqslant r \leqslant \infty$, can be joined by a C^r path $f:[0,1] \to M$, and for $r \geqslant 1$, f can be chosen to be an embedding. (This is also true, but difficult, for $r = \omega$.)

3. Let $p: M \to N$ be a C^s map and $f: N \to M$ a C^r section of p (that is, $pf = 1_M$).

(a) If $1 \leqslant r < s \leqslant \infty$ then f can be C^r approximated by C^s sections. In fact if g is sufficiently C^r close to f then $g(M)$ is the image of a section.

(b) If $0 = r < s \leqslant \infty$ and p is submersive, then f can be C^r approximated by C^s sections.

4. There are relative versions of theorems 2.6, 2.9, 2.12, and 2.13.

5. What are the bilinear functions A_k in the proof of Theorem 2.4?

3. Approximations on ∂-Manifolds and Manifold pairs

In this section we extend the approximation results of Section 2.2 to ∂-manifolds and manifold pairs.

In constructing a C^∞ differential structure for a C^r ∂-manifold, one needs to know that a C^r map $f:(M,\partial M) \to (N,\partial N)$ can be approximated by C^∞ maps. Thus we are led to consider the space $C_S^r(M,\partial M; N,\partial N)$ of C^r maps $f: M \to N$ such that $f(\partial M) \subset \partial N$, with the strong topology. More generally we consider $C_S^r(M,M_0; N,N_0)$, where $M_0 \subset M$ and $N_0 \subset N$ are closed neat C^r submanifolds.

The proofs are quite close to those of Section 2.2. The main change is an adaptation of the approximation Theorem 2.4 to pairs. The details of this are given for ∂-manifolds; other proofs are omitted.

The definition of $\|f\|_{r,K}$, where now f is defined on an open subset of a halfspace, is the same as before.

3.1. Lemma. *Let $E \subset \mathbb{R}^m$ and $F \subset \mathbb{R}^n$ be halfspaces, $U \subset E$ an open subset, and $f: U \to F$ a C^r map, $0 \leqslant r < \infty$. Let $K \subset U$ be compact and $\varepsilon > 0$. There is an open neighborhood $U' \subset U$ of K and a C^∞ map $g: U' \to F$ such that $\|g - f\|_{r,U'} < \varepsilon$. Moreover, if $X \subset \partial U$ is such that $f(X) \subset \partial F$, then g can be chosen so that $g(X \cap U') \subset \partial F$.*

Proof. We may assume that either ∂U or ∂F is nonempty, for if both are empty Theorem 3.1 is subsumed by previous results. If $\partial U = \varnothing$, first approximate f by a map

$$f_0(x) = f(x) + y$$

where $y \in F - \partial F$ has norm $< \varepsilon$; then apply 2.4.

If $\partial U \neq \varnothing$ but $\partial F = \varnothing$, extend f to a C^r map on an open neighborhood of V in $\mathbb{R}^m$, using local extensions and C^r partitions of unity, and apply 2.4.

If both ∂U and ∂F are nonempty, make the natural identifications:

$$\mathbb{R}^m = (\partial E) \times \mathbb{R}, \qquad E = (\partial E) \times [0,\infty);$$
$$\mathbb{R}^n = (\partial F) \times \mathbb{R}, \qquad F = (\partial F) \times [0,\infty).$$

For $(x,y) \in (\partial E) \times \mathbb{R}$ write

$$f(x,y) = (f_0(x,y), f_1(x,y)) \in \partial F \times [0,\infty).$$

Note that $f_1 \geqslant 0$.

We use a variation of the convolution of Section 2.1. Let $\theta:\mathbb{R}^m \to R$ be a C^∞ convolution kernel of the special form

$$\theta(x,y) = \alpha(x)\beta(y)$$

where $\alpha:\partial E \to R$, $\beta:R \to R$ are C^∞ convolution kernels.

Suppose α, β and θ have support radius less then $\delta > 0$. Let $U' = \{(x,y) \in U | (x,z) \in U$ if $y \leqslant z < y + \delta$. Define

$$h: U' \to \mathbb{R}^n,$$

$$h(x,y) = \int_{t\geqslant 0} \int_{s\in\partial E} f(x - x, y + t)\alpha(s)\beta(t)\, ds\, dt$$

Then h is C^∞ and $\|h - f\|_{r,U'} \to 0$ as $\delta \to 0$. Moreover $h(U') \subset F$ because f_1 and β are nonnegative. If $f(\partial U) \subset \partial F$ then $f(x,y) \geqslant f(x,0)$, which implies $h(x,y) \geqslant h(x,0)$. Now define $g(x,y) = h(x,y) - h(x,0)$. Then $g(U') \subset F$ and $g(\partial U') \subset \partial F$. If δ is small enough, the map $g: U' \to F$ satisfies the lemma.

3.2. Lemma. *Let $E \subset \mathbb{R}^m$, $F \subset \mathbb{R}^n$ be halfspaces and $U \subset E$, $V \subset F$ open sets. Then $C^\infty(U,V)$ is dense in $C^r_S(U,V)$, and $C^\infty(U,\partial U; V,\partial V)$ is dense in $C^r_S(U,\partial U; V,\partial V)$, $0 \leqslant r < \infty$.*

Proof. This proof is almost the same as that of Theorem 2.4. The details are left to the reader.

3.3. Theorem. *Let M and N be a C^s manifolds, $1 \leqslant s \leqslant \infty$; ∂M or ∂N or both may be nonempty. Then $C^s(M,N)$ is dense in $C^r_S(M,N)$ and $C^s(M,\partial M; N,\partial N)$ is dense in $C^r_S(M,\partial M; N,\partial N)$, $0 \leqslant r < s$.*

Proof. A relative version of Theorem 3.2 is proved in the same way as Theorem 2.5. The globalization to Theorem 3.3 is just like the proof of Theorem 2.6.

QED

3.4. Theorem. *Every C^r manifold M, $1 \leqslant r < \infty$, is C^r diffeomorphic to a C^∞ manifold and the latter is unique up to C^∞ diffeomorphism.*

Proof. Similar to the proof of Theorem 2.9, and left to the reader.

By a *C^r manifold pair* (M,M_0) we mean a C^r manifold M together with a C^r submanifold M_0. The approximation and globalization techniques developed so far can be combined to yield the following results; the proofs have the same general outline as the previous ones and are left to the reader.

3.5. Theorem. *Let (M,M_0) and (N,N_0) be C^s manifold pairs, $1 \leqslant s \leqslant \infty$. Suppose that M_0 is closed in M, and $M_0 \subset M - \partial M$ or $M_0 \subset \partial M$ or M_0 is a neat submanifold. Then $C^s(M,M_0; N,N_0)$ is dense in $C^r(M,M_0; N,N_0)$, $0 \leqslant r < s$. If $1 \leqslant r < s$, and (M,M_0) and (N,N_0) are C^r diffeomorphic, they are also C^s diffeomorphic.*

3.6. Theorem. *Let (M,M_0) be a C^r manifold pair. If $0 < r < s \leqslant \infty$ then (M,M_0) has a compatible C^s structure (that is, (M,M_0) is C^r diffeomorphic to a C^s manifold pair). If also M_0 is closed in M, and $M_0 \subset M - \partial M$ or $M_0 \subset \partial M$ or M_0 is neat, then the compatible C^s structure is unique up to C^r diffeomorphism of manifold pairs.*

Theorem 3.6 is of use in parts of analysis (invariant manifold theory, for example) where submanifolds of low differentiability occur naturally.

There are counter-examples to the existence of C^∞ structures on C^0 pairs (M,M_0), even where M and M_0 each have C^∞ structures.

We leave to the reader the adaptation of the proofs of Theorems 2.12, 2.13 and 2.14 to ∂-manifolds.

Exercises

1. Let $1 \leqslant r < s \leqslant \infty$. There are C^s manifolds, M, N and closed sets $A \subset M$, $B \subset N$ such that $C^s(M,A; N,B)$ is *not* dense in $C^r_S(M,A; N,B)$. [Hint: let $A = M$. Suppose $B \subset N$ is a C^r submanifold which is not C^s, and $f: M \to B$ is a C^r diffeomorphism.]

2. Relative versions of Theorems 3.3 through 3.6 are true.

3. Theorems 3.5 and 3.6 extend to maps of *manifold n-ads* $\{M_i\} \to \{N_i\}$ where $M_n \subset \cdots \subset M_0 \subset M$ and $N_n \subset \cdots \subset N_0 \subset N$ are nested families of closed neat submanifolds.

4. Let M be a C^∞ manifold and $A \subset M$ a closed neat submanifold. If $q > 2 \dim M$ then every C^∞ embedding of ∂M in $\mathbb{R}^q$, or of A in $\mathbb{R}^q$, extends to C^∞ embedding of M.

4. Jets and the Baire Property

It is convenient to redefine the topologies on $C^r(M,N)$ in a way which avoids coordinate charts. $C^r(M,N)$ will be identified with a subset of $C^0(M,J^r(M,N))$ where $J^r(M,N)$ is the manifold of r-jets of maps from M to N. In this way $C^r(M,N)$ becomes a set of continuous maps. Our first goal in this section is to define the weak and strong topologies on such sets.

We denote by $C(X,Y)$ the set of continuous maps from a space X to a space Y. The *compact open* topology on $C(X,Y)$ is generated by the subbase comprising all sets of the form

$$\{f \in C(X,Y): f(K) \subset V\}$$

where $K \subset X$ is compact and $V \subset Y$ is open. We also call this the *weak* topology to contrast it with another topology defined below. The resulting topological space is denoted by $C_W(X,Y)$.

The weak topology is most useful when X is locally compact. When Y is a metric space the topology is the same as that of uniform convergence on compact sets. If X is compact and Y is metric, $C_W(X,Y)$ has the metric

$$d(f,g) = \sup_x d(f(x),g(x)).$$

This metric is complete provided Y is a complete metric space. More generally:

4.1. Theorem. *Let each component of X be locally compact with a countable base; let Y be a complete metric space. Then $C_W(X,Y)$ has a complete metric.*

Proof. It suffices to construct a complete metric on $C_W(X_\alpha Y)$ for each component X_α of X; therefore we assume X locally compact with a countable base. Then X has a countable covering by compact sets $\{X_n\}$. Each space $C_W(X_n, Y)$ has a complete metric.

Define a map

$$\rho: C_W(X, Y) \to \prod_n C_W(X_n, Y),$$
$$\rho_n(f) = f|X_n.$$

Then ρ is a homeomorphism onto a closed subspace. Since the product of a countable number of complete metric spaces has a complete metric, $C_W(X,Y)$ is homeomorphic to a closed subspace of a complete metric space and thus has a complete metric.

QED

Now let X and Y be arbitrary spaces. The space $C_S(X,Y)$ is the set $C(X,Y)$ with the following *strong topology*. Let $\Gamma_f \subset X \times Y$ denote the graph of the map f. If $W \subset X \times Y$ is an open set containing Γ_f, let

$$\mathcal{N}(f,W) = \{g \in C(X,Y): \Gamma_g \subset W\}.$$

These sets, for all f and W, form a base for the strong topology. The induced topology on a subset of $C(X,Y)$ is also called *strong*.

When X is paracompact and Y is metric, $C_S(X,Y)$ has the base comprising all sets of the form

$$\mathcal{N}(f,\varepsilon) = \{g : d(g(x), f(x)) < \varepsilon(x), \quad \text{all} \quad x \in X\}$$

where $f \in C(X,Y)$ and $\varepsilon \in C(X,R_+)$ are arbitrary.

If X is compact the weak and strong topologies are the same.

We cannot expect the strong topology to have a complete metric, since it may not have any metric. But we shall see that in many cases it is a *Baire space*, that is, the intersection of a countable family of dense open sets is dense.

Let Y be a metric space. A subset of $C(X,Y)$ is *uniformly closed* if it contains the limit of every uniformly convergent sequence in it. Observe that this concept depends on the metric on Y. A subset which is closed under pointwise convergence is uniformly closed, as is a subset which is closed in the weak topology.

4.2. Theorem. *Let X be a paracompact space and Y a complete metric space. Then every uniformly closed subset $Q \subset C(X,Y)$ is a Baire space in the strong topology.*

Corollary. *If M and N are C^0 manifolds, every weakly closed subset of $Q \subset C(M,N)$ is a Baire space in the strong topology.*

Proof. Let $\{A_n\}_{n \in N}$ be a sequence of dense open subsets of Q (referring always to the strong topology), and let $U \subset Q$ be a nonempty open set. Then $A_0 \subset U$ is a nonempty open subset of Q. Therefore there exists $f_0 \in A_0 \cap U$ and $\varepsilon_0 \in C(X,\mathbb{R}_+)$ such that

$$Q \cap \mathcal{N}(f_0,\varepsilon_0) \subset A_0 \cap U$$

where $\mathcal{N}(f_0,\varepsilon_0) = \{g : d(f_0x,gx) \leqslant \varepsilon(x)\}$. We may obviously assume $\varepsilon_0 < 1$.

By recursion there are sequences $\{f_n\}$ in Q and $\{\varepsilon_n\}$ in $C(X,\mathbb{R}_+)$ such that for all $n \in N$:

$$Q \cap \mathcal{N}(f_{n+1},\varepsilon_{n+1}) \subset A_{n+1} \cap \mathcal{N}(f_n,\varepsilon_n),$$

and $\varepsilon_{n+1} \leqslant \varepsilon_n/2$. The sequence $\{f_n\}$ satisfies

$$d(f_{n+1}x,f_nx) \leqslant 2^{-n}$$

and so is uniformly convergent. The limit f is in Q since Q is uniformly closed. Also f belongs to every $\mathcal{N}(f_n,\varepsilon_n)$, so $f \in U$ and also $f \in \cap A_n$.

QED

We now define jets of finite order r, treating first manifolds without boundary. Let M, N be C^r manifolds, $0 \leqslant r < \infty$. An *r-jet from M to N* is an equivalence class $[x,f,U]_r$ of triples (x,f,U), where $U \subset M$ is an open set, $x \in U$, and $f : U \to N$ is a C^r map; the equivalence relation is: $[x,f,U]_r = [x',f',U']_r$ if $x = x'$ and in some (and hence any) pair of charts adapted to f at x, f and f' have the same derivatives up to order r. We use the notation

$$[x,f,U]_r = j_x^r f = j^r f(x)$$

to denote the *r-jet of f at x*. We call x the *source* and $f(x)$ the *target* of $[x,f,U]$.

The set of all r-jets from M to N is denoted by $J^r(M,N)$. There are well defined *source* and *target maps*:

$$\sigma : J^r(M,N) \to M, \qquad \sigma[x,f,U]_r = x$$
$$\tau : J^r(M,N) \to N, \qquad \tau[x,f,U]_r = f(x).$$

We put

$$\sigma^{-1}(x) = J_x^r(M,N), \qquad \tau^{-1}(y) = J^r(M,N)_y$$

and

$$J_x^r(M,N) \cap J^r(M,N)_y = J_{x,y}^r(M,N);$$

this last is the set of all r-jets from M to N with source x and target y.

Consider the special case $M = \mathbb{R}^m$, $N = \mathbb{R}^n$. We write

$$J^r(\mathbb{R}^m,\mathbb{R}^n) = J^r(m,n).$$

Suppose $U \subset \mathbb{R}^m$ is open and $f \in C^r(U,\mathbb{R}^n)$. The r-jet of f at $x \in U$ has a canonical representative, namely the Taylor polynomial of f of order r at x. This polynomial map from $\mathbb{R}^m$ to $\mathbb{R}^n$ is uniquely determined by the list of derivatives of order r of f at x. This list belongs to the vector space

$$P^r(m,n) = \mathbb{R}^n \times \prod_{k=1}^{r} L^k_{\text{sym}}(\mathbb{R}^m,\mathbb{R}^n)$$

where $L^k_{\text{sym}}(\mathbb{R}^m,\mathbb{R}^n)$ denotes the vector space of symmetric k-linear maps from $\mathbb{R}^m$ to $\mathbb{R}^n$. Conversely any element of $P^r(m,n)$ comes from a unique jet in $J^r_x(m,n)$. In this way we have identifications

$$J^r_x(m,n) = P^r(m,n)$$

and

$$J^r(m,n) = \mathbb{R}^m \times P^r(m,n).$$

In particular $J^r(m,n)$ is a finite dimensional vector space (for r finite). If $U \subset \mathbb{R}^m$ and $V \subset \mathbb{R}^n$ are open sets then $J^r(U,V)$ is an open subset of $J^r(m,n)$.

Now let M, N be manifolds of dimension m, n respectively. Suppose at first that ∂M and ∂N are empty. If (φ,U), (ψ,V) are charts for M, N the following map $\theta: J^r(U,V) \to J^r(\varphi U,\psi V)$ is a bijection:

$$\theta: j^r_x f \mapsto j^r_y(\psi f \varphi^{-1}), \qquad y = f(x).$$

Thus θ sends each jet to the jet of its local representation. Now $J^r(\varphi U,\psi V)$ is an open set in the vector space $J^r(m,n)$, which is isomorphic to a Euclidean space. Therefore we can view $(\theta,J^r(U,V))$ as a chart on $J^r(M,N)$; the topology on $J^r(M,N)$ is of course that determined by these charts. In this way $J^r(M,N)$ is a C^0 manifold. In fact if M, N are C^{r+s} manifolds, $J^r(M,N)$ has differentiability class C^s.

For each C^r map $f: M \to N$ we define a map

$$j^r f: M \to J^r(M,N)$$

by $x \mapsto j^r f(x)$. This *r-prolongation* of f is continuous and in fact C^s if M and N are C^{r+s}. We consider $j^r f$ as a kind of intrinsic r'th derivative of f. It is clear that j^r is injective.

4.3. Theorem. *The image of*

$$j^r: C^r(M,N) \to C^0(M,J^r(M,N))$$

is closed in the weak topology.

Proof. We must show that the image is closed under uniform convergence on compact sets. It suffices to consider convex compact subsets of coordinate charts. Ultimately we must prove that if $U \subset \mathbb{R}^m$ is open and $\{f_n\}$ is a sequence such that for each $k = 0, \ldots, r$ the sequence $\{D^k f_n(x)\}$ converges uniformly on U to a continuous map $g_k: U \to L^k(\mathbb{R}^m,\mathbb{R}^n)$, then $g_k = D^k g_0$. This is proved by induction on k. The inductive step is the same as the case $k = 1$. If Df_n converges uniformly to g_1 and f_n converges uniformly to g_0, we have, for $x, x + y \in U$:

$$\begin{aligned} g_0(x + y) &= \lim_{n\to\infty} f_n(x + y) \\ &= \lim_{n\to\infty} f_n(x) + \lim_{n\to\infty} \int_0^1 Df_n(x + ty)y\,dt \\ &= g_0(x) + \int_0^1 g_1(x + ty)y\,dt, \end{aligned}$$

by uniform convergence. It follows easily that $g_1 = Dg_0$.

QED

If we give $C^r(M,N)$ the topology induced by j^r from the weak or strong topology on $C^0(M,J^r(M,N))$, we obtain spaces which coincide with $C^r_W(M,N)$ and $C^r_S(M,N)$, as the reader can verify.

From Theorems 4.1, 4.2, 4.3 we obtain:

4.4. Theorem.

(a) *$C^r_W(M,N)$ has a complete metric;*

(b) *Every weakly closed subspace of $C^r_S(M,N)$ is a Baire space (in the strong topology).*

Suppose M and N are C^∞ manifolds. We define the set $J^\infty(M,N)$ to be the inverse limit of the sequence

$$J^0(M,N) \leftarrow J^1(M,N) \leftarrow \cdots$$

and $J^\infty_x(M,N)$ to be the inverse limit of the sequence

$$J^0_x(M,N) \leftarrow J^1_x(M,N) \leftarrow \cdots$$

An element of $J^\infty_x(M,N)$ is, *by definition*, an ∞-jet at x.

The maps j^r fit together to define a map

$$j^\infty : C^\infty(M,N) \to C^0(M,J^\infty(M,N)).$$

Again the image is weakly closed, and the weak and strong topologies on $C^\infty(M,N)$ are the same as those induced by j^∞ from the corresponding topologies on $C(M,J^\infty(M,N))$. It follows that $C^\infty_W(M,N)$ has a complete metric and every weakly closed subspace of $C^\infty_S(M,N)$ is Baire in the strong topology. In particular, $C^\infty_S(M,N)$ is a Baire space.

Returning to the density of embeddings, we give an alternative proof of Theorem 2.13. It suffices to prove that if $f_0: M \to N$ is a C^r proper map and $\mathcal{N} \subset \operatorname{Prop}^r(M,N)$ is a neighborhood of f_0 then $\mathcal{N}$ contains an injective immersion, for a proper injective immersion is an embedding.

We may assume that

$$\mathcal{N} = \mathcal{N}(\Phi,\Psi,K,\varepsilon),$$

the notation being as usual, where $K = \{K_i\}_{i\in\Lambda}$ is a family of coordinate disks which covers M.

For each $i \in \Lambda$ let

$$X_i = \{f \in \mathcal{N} : f|K_i \text{ is an embedding}\}.$$

Then X_i is dense and open in $\mathcal{N}$. To see this let $B_i \subset U_i$ be a slightly larger coordinate disk containing K_i in its interior. By Theorem 1.3.5 we can approximate $f|B_i$ by an embedding g_i; glueing g_i and f together by a C^r map $\lambda: M \to [0,1]$ which is 1 on D_i and 0 on $M - K_i$, gives a map which embeds D_i and which tends to f as g_i tends to $f|B_i$. Thus X_i is dense. Openness follows from openness of embeddings.

A similar argument proves that if (i,j) is such that $K_i \cap K_j = \varnothing$, then the set

$$X_{ij} = \{f \in \mathcal{N} : f|K_i \cup K_j \text{ is an embedding}\}$$

is dense and open.

Let $K^{(1)}, K^{(2)}, \ldots$ be a family of refinements of K such that each $K^{(j)}$ is a locally finite covering of M by coordinate disks, and such that for any distinct $x, y \in M$ there exist disjoint disks $K_1^{(j)}, K_2^{(j)} \in K^{(j)}$ with $x \in K_1^{(j)}$, $y \in K_2^{(j)}$. Since M has a countable base, each $K^{(n)}$ is countable.

Let $X^{(n)}$ be the set of $f \in \mathcal{N}$ such that $f|K_i^{(n)} \cup K_j^{(n)}$ is an embedding whenever $K_i^{(n)}, K_j^{(n)}$ are disjoint disks in $K^{(n)}$. Then each $X^{(n)}$, and hence $\bigcap_n X^{(n)}$, is the intersection of a countable family of dense open subsets of $\mathcal{N}$. Since the Baire property is inherited by open sets, $\mathcal{N}$ is a Baire space. Therefore $\bigcap_n X^{(n)}$ is dense in $\mathcal{N}$. This intersection is precisely the set of injective immersions in $\mathcal{N}$. Therefore embeddings are dense.

In our treatment of jets we have assumed that $\partial M = \partial N = \varnothing$. We now consider the general case where M and N are allowed to have boundaries.

The definition of r-jet is unchanged, but the topology on $J^r(M,N)$ must be treated carefully. Consider first open subsets U, V of halfspaces $E \subset \mathbb{R}^m$, $F \subset \mathbb{R}^n$. For each $(x,y) \in U \times V$ there are canonical identifications (for $r < \infty$):

$$J^r_{x,y}(U,V) = \prod_{k=1}^{r} L^k_{\text{sym}}(\mathbb{R}^m,\mathbb{R}^n)$$
$$= J^r_{0,0}(m,n).$$

Consequently

$$J^r(U,V) = U \times V \times J^r_{0,0}(m,n).$$

If either ∂U or ∂V is empty, this is an open subset of a halfspace. But if $\partial U \neq \varnothing$ and $\partial V \neq \varnothing$ it is not. It is, however, homeomorphic to an open subset of a half space; this follows from the same property for $U \times V$, which in turn follows from the homeomorphism

$$[0,\infty) \times [0,\infty) \approx \mathbb{R} \times [0,\infty).$$

Thus again $J^r(M,N)$ has a natural C^0 manifold structure, and the preceding development goes through. (But if M and N are C^{r+s} ∂-manifolds, $J^r(M,N)$ has no natural C^s structure.) The treatment of jets of infinite order is the same as before, and Theorem 3.4 holds for all manifolds. The proof of density of embeddings in $\text{Prop}_S^r(M,N)$ can be adapted to ∂-manifolds.

Exercises

1. Let X be locally compact and paracompact. Suppose Y has an open covering by completely metrizable subspaces. Then every weakly closed subset of $C_S(X,Y)$ is Baire in the strong topology.

*****2.** Under what conditions is the natural map

$$C_S(X,C_S(Y,Z)) \to C_S(X \times Y,Z)$$

a homeomorphism?

3. Let X be paracompact and Y metric. For each $\varepsilon \in C(X,R_+)$ define a metric d_ε on $C(X,Y)$:

$$d_\varepsilon(f,g) = \min\{1, \sup_x d(fx,gx)/\varepsilon(x)\}.$$

(a) If Y is complete each metric d_ε is complete.

(b) If $Q \subset C(X,Y)$ is uniformly closed and ε is bounded then Q is closed in the metric space $(C(X,Y),d_\varepsilon)$.

(c) The strong topology on $C(X,Y)$ is that induced by the family of metrics

$$\{d_\varepsilon : \varepsilon \in C(X,R_+)\}.$$

4. The Baire property for the strong topology on uniformly closed subsets of $C(X,Y)$ follows from Exercise 3 and the following. Let Q be a space whose topology is defined by a family Δ of complete metrics. Suppose that Δ is a directed set under the partial ordering:

$$d_1 \leqslant d_2 \quad \text{if} \quad d_1(x,y) \leqslant d_2(x,y) \quad \text{for all} \quad (x,y).$$

Then Q has the Baire property.

5. Let $\{X_i\}_{i\in A}$ be a family of complete metric spaces. Let X be the product of the sets X_i, with the following *strong product topology*: a set is open if and only if its projection into each X_i is open. If $Q \subset X$ is closed in the usual product topology, the strong product topology on Q has the Baire property.

6. If M is compact then $C_W^r(M,\mathbb{R}^n)$ is a Banach space for $r < \infty$.

***7.** $C_W^r(\mathbb{R},\mathbb{R})$ is a complete, locally convex topological vector space, but it does not have a norm. Thus it is not a Banach space. [Hint: let E be a topological vector space. Call $X \subset E$ *bounded* if for every neighborhood $N \subset E$ of 0 there exists $t > 0$ such that $tX \subset N$. Then E has a norm if and only if there is a bounded convex neighborhood of 0.]

8. $C_W^\infty(M,\mathbb{R})$ is a separable, complete locally convex topological vector space which is separable if M is compact; but it does not have a norm. (See Exercise 8.)

9. Let M be a C^r manifold, $0 \leqslant r \leqslant \omega$. The set $\text{Diff}^r(M)$ of C^r diffeomorphisms of M is a topological group under composition in both the weak and the strong topologies.

10. Let M, N, P be C^r manifolds, $0 \leqslant r \leqslant \infty$.

(a) The composition map

$$C^r(N,P) \times C^r(M,N) \to C^r(M,P),$$
$$(f,g) \to f \circ g$$

is continuous in the weak topologies.

(b) For fixed $f = f_0$ composition is continuous in g in the strong topologies.

(c) For fixed $g = g_0$ composition is continuous in f in the strong topologies if and only if g is proper.

(d) Composition is continuous at (f_0, g_0) if and only if g_0 is proper.

11. Compute the dimension of $J^r(M,N)$.

******12.*** Is $C^0_S(\mathbb{R},\mathbb{R})$ paracompact? Normal?

13. The subspace

$$X = \{f \in C^0_S(\mathbb{R},\mathbb{R}) : \text{Supp } f \text{ is compact}\}$$

is closed but does not have the Baire property.

14. The *limit set* $L(f)$ of $f: M \to N$ is the set of $y \in N$ such that $y = \lim x_n$ for some sequence $\{x_n\}$ in M which has no convergent subsequence. If $\dim N > 2 \dim M$ and $1 \leqslant r \leqslant \infty$ then embeddings are dense in

$$\mathscr{L} = \{f \in C^r_S(M,N) : f(M) \cap L(f) = \varnothing\}.$$

[If $f \in \mathscr{L}$, there is an open set $N_0 \subset N$ containing $f(M)$ such that $f: M \to N_0$ is proper; use Theorem 2.13.]

15. An open set $P \subset J^r(m,n)$ is *natural* if it is closed under composition with jets of local diffeomorphisms of $\mathbb{R}^m$ and $\mathbb{R}^n$. Given such a P and C^r manifolds M, N define $P(M,N)$ to be the set of C^r maps from M to N all of whose local representations have their r-jets in P. Then $P(M,N)$ is open in $C^r_S(M,N)$.

16. The set of immersions is a Baire subset of $C^r_W(M,N)$, $1 \leqslant r \leqslant \infty$ if $\dim M \leqslant 2 \dim N$. (A *Baire* subset is the intersection of a countable family of dense open subsets.)

***17.** The space $C^r_S(M,N)$ is completely regular, $0 \leqslant r \leqslant \infty$.

5. Analytic Approximations

Partitions of unity are of no use for constructing analytic approximations because an analytic function on $\mathbb{R}^n$ is constant if it has bounded support. More subtle globalization techniques are needed.

Using methods from complex analysis, Grauert and Remmert [1] have proved the following deep result:

5.1. Theorem. *Let M and N be C^ω manifolds. Then $C^\omega(M,N)$ is dense in $C^r_S(M,N)$, $0 \leqslant r \leqslant \infty$.*

That this is true is very fortunate, for it means that C^ω differential topology is no different from the C^∞ theory for such questions as diffeomorphism classification of manifolds, existence of embeddings and immersions, etc. These questions concern *open* sets of maps, from one manifold to another. Where *closed* sets of maps or individual maps are considered, for example solutions to differential equations, the degree of differentiability may play an important role. It is also an important consideration whenever maps from a manifold to itself are studied. For such maps problems of conjugacy and iteration arise, and high differentiability is sometimes a crucial hypothesis.

Occasionally analytic maps are a useful tool because their level surfaces—even critical ones—are analytic varieties. More general than manifolds, analytic varieties still are topologically fairly simple. In particular, they can be triangulated.

Morrey [1] proved Theorem 5.1 for compact manifolds. An elegant treatment of this case was given by Bochner [1] under the assumption of an analytic Riemannian metric.

Once Theorem 5.1 is known, the existence of compatible C^ω structures on C^r manifolds can be proved using the maximality argument of Theorem 2.9. This was first proved by Whitney [2] using the following easier approximation result:

5.2. Theorem (Whitney [6]) *Let $U \subset \mathbb{R}^m$ be open and $f: U \to \mathbb{R}$ a C^r map, $0 \leqslant r < \infty$. Let $v: \mathbb{R}^m \to [0,1]$ be a C^∞ map of bounded support, equal to 1 on a neighborhood of a compact set $K \subset U$. Set $h(x) = v(x)f(x)$. Let $\delta: \mathbb{R}^m \to \mathbb{R}$, $\delta(x) = \exp(-|x|^2)$. Let $T = 1/\int_{\mathbb{R}^n} \delta$. Let $\varepsilon > 0$. Then for $\kappa > 0$ sufficiently large, the convolution g of h with the function $T\kappa^m \, \delta(\kappa,x)$ is analytic and satisfies $\|g - f\|_{r,K} < \varepsilon$.*

The proof is straightforward, but in the absence of partitions of unity it is not easy to pass from Theorem 5.2 to an approximation theorem for abstract C^ω manifolds. For C^ω submanifolds of $\mathbb{R}^n$, however, Theorem 5.2 works quite nicely once the technique of tubular neighborhoods is available.

We shall use Theorem 5.2 to prove, in Section 4.6, that C^∞ manifolds have compatible C^ω structures.

Nash [1] proved that a compact connected C^∞ manifold without boundary is C^∞ diffeomorphic to a component of a real algebraic variety; he also proved an algebraic approximation theorem for maps of such manifolds.

An interesting topological application of Nash's results was made by Artin and Mazur [1]: if M is a compact connected C^∞ manifold there is a dense subset of $C_S^\infty(M,M)$ of maps $f: M \to M$ such that the number of fixed points of f^n is bounded above by a function of the form $Ae^{\lambda n}$ where A and λ are positive constants depending only on M.

Exercise

1. If $U \subset \mathbb{R}^m$ is open, polynomials are dense in $C_W^\infty(U,\mathbb{R})$. [Hint: replace the exponential in Theorem 5.2 by a Taylor polynomial.]

Chapter 3

Transversality

Transversality unlocks the secrets of the manifold.

—H. E. Winkelnkemper

"Transversal" is a noun; the adjective is "transverse."

—J. H. C. Whitehead, 1959

Consider the following statements:

1. If $f: S^2 \to \mathbb{R}^2$ is C^1, then $f^{-1}(y)$ is finite for "most" points $y \in \mathbb{R}^2$.
2. Two lines in $\mathbb{R}^3$ do not intersect "in general."
3. If $f: \mathbb{R} \to \mathbb{R}$ is C^1, "almost all" horizontal lines in $\mathbb{R} \times \mathbb{R}$ are nowhere tangent to the graph of f.
4. "Generically" a C^1 immersion $S^1 \to \mathbb{R}^2$ has only a finite number of crossing points.

These statements illustrate a type of reasoning that is common in differential topology. Most people would agree they are plausible. Yet there is an element of uncertainty about them, due to the vagueness of the words in quotation marks. Even if these are given precise definitions, it is obvious that something needs to be proved. The purpose of this chapter is to develop the mathematics needed to justify such statements.

The basis of this mathematics is a profound result in analysis, due to A. P. Morse and A. Sard.[1] It says that if $f: \mathbb{R}^n \to \mathbb{R}^k$ is C^r, where $r > \max\{0, n - k\}$, then the set of critical values has measure zero in $\mathbb{R}^k$. We prove this only for the case $r = \infty$. This version is considerably easier and is adequate for differential topology.

The reader may prefer to accept the Morse–Sard theorem (Theorem 1.3) on faith, since the method of proof is not used elsewhere.

In Section 3.2 the Morse–Sard theorem is used to prove various transversality theorems. These guarantee the existence of plenty of maps $f: M \to N$ which are transverse to a submanifold $A \subset N$. This is a fundamental result in differential topology; analogous statements in the theory of topological or polyhedral manifolds are false.

In this chapter and the remainder of the book, *all manifolds and submanifolds are assumed to be C^∞* unless the contrary is stated. In view of the approximation results of the preceding chapter this is not a serious restriction.

---

[1] The first theorem of this kind was proved by A. B. Brown [1]. See also Dubovickiĭ [1].

1. The Morse–Sard Theorem

An *n-cube* $C \subset \mathbb{R}^n$ of *edge* $\lambda > 0$ is a product

$$C = I_1 \times \cdots \times I_n \subset \mathbb{R} \times \cdots \times \mathbb{R} = \mathbb{R}^n$$

of closed intervals of length λ; thus

$$I_j = [a_j, a_j + \lambda] \subset \mathbb{R}.$$

The *measure* (or *n-measure*) of C is

$$\mu(C) = \mu_n(C) = \lambda^n.$$

A subset $X \subset \mathbb{R}^n$ has *measure zero* if for every $\varepsilon > 0$ it can be covered by a family of n-cubes, the sum of whose measures is less than ε. A countable union of sets of measure zero has measure zero. Therefore X has measure zero if every point of X has a neighborhood in X of measure zero (by Lindelöf's principle).

1.1. Lemma. *Let $U \subset \mathbb{R}^n$ be an open set and $f: U \to \mathbb{R}^n$ a C^1 map. If $X \subset U$ has measure zero, so has $f(X)$.*

Proof. Every point of X belongs to an open ball $B \subset U$ such that $\|Df(x)\|$ is uniformly bounded on B, say by $\kappa > 0$. Then

$$|f(x) - f(y)| \leqslant \kappa|x - y|$$

for all $x, y \in B$. It follows that if $C \subset B$ is an n-cube of edge λ, then $f(C)$ is contained in an n-cube C' of edge less than $\sqrt{n}\kappa\lambda = L\lambda$. Therefore $\mu(C') < L^n\mu(C)$.

Write $X = \bigcup_1^\infty X_j$ where each X_j is a compact subset of a ball B as above. For each $\varepsilon > 0$, $X_j \subset \bigcup_k C_k$ where each C_k is an n-cube and $\sum \mu(C_k) < \varepsilon$. It follows that $f(X_j) \subset \bigcup_k C'_k$ where the sum of the measures of the n-cubes C'_k is less than $L^n\varepsilon$. Hence each $f(X_j)$ has measure zero, and so X has measure zero.

QED

Now let M be a (C^∞) n-dimensional manifold. A subset $X \subset M$ is said to have *measure zero* if for every chart (φ, U), the set $\varphi(U \cap X) \subset \mathbb{R}^n$ has measure zero. Because of Lemma 1.1, this will be true provided there is *some* atlas of charts with this property.

Notice that we have not defined the "measure" of a subset of M, but only a certain kind of subset which we say "has measure zero." This is in accordance with the red herring principle (Chapter 1, page 22, footnote).

It can be shown that a cube does not have measure zero. Therefore a set of measure zero in $\mathbb{R}^n$ cannot contain a cube; hence it has empty interior. It follows that a *closed* measure zero subset of $\mathbb{R}^n$, or of a manifold M, is nowhere dense. More generally, suppose $X \subset M$ has measure zero and is

σ-compact, that is, X is the union of a countable collection of compact sets. Each of these is nowhere dense, and so X is nowhere dense by the Baire category theorem. The complement of X is *residual*, that is, it contains the intersection of a countable family of dense open sets. The Baire theorem says a residual subset of a complete metric space is dense. Note that the intersection of countably many residual sets is residual.

1.2. Proposition. *Let M, N be manifolds with* $\dim M < \dim N$. *If $f: M \to N$ is a C^1 map then $f(M)$ is nowhere dense.*

Proof. It suffices to show that $f(M)$ has measure zero. This follows from: $g(U) \subset \mathbb{R}^n$ has measure zero if $U \subset \mathbb{R}^m$ is open and $g: U \to \mathbb{R}^n$ is C^1, with $m < n$. To prove this assertion, write g as a composition of C^1 maps

$$U = U \times 0 \subset U \times \mathbb{R}^{n-m} \xrightarrow{\pi} U \xrightarrow{g} \mathbb{R}^n.$$

Clearly $U \times 0$ has n-measure zero in

$$U \times \mathbb{R}^{n-m} \subset \mathbb{R}^m \times \mathbb{R}^{n-m} = \mathbb{R}^n;$$

hence the proposition follows from Lemma 1.1 applied to πg.

QED

Recall that a point $x \in M$ is *critical* for a C^1 map $f: M \to N$ if the linear map $T_x f: M_x \to N_{f(x)}$ is not surjective. We denote by $\sum_f$ the set of critical points of f. Note that $N - f(\sum_f)$ is the set of regular values of f.

1.3. Morse–Sard Theorem. *Let M, N be manifolds of dimensions m, n and $f: M \to N$ a C^r map. If*

$$r > \max\{0, m - n\}$$

then $f(\sum_f)$ has measure zero in N. The set of regular values of f is residual and therefore dense.

The differentiability requirement is strange but necessary. We shall prove the theorem only in the C^∞ case. Before beginning the proof let us examine the implications of the theorem in particular instances.

Let $f: \mathbb{R} \to \mathbb{R}$ be C^1. If y is a regular value, then the horizontal line $R \times y \subset \mathbb{R} \times \mathbb{R}$ is transverse to the graph of f (Figure 3–1). Thus the theorem implies that "most" horizontal lines are transverse to the graph.

Consider next a map $f: \mathbb{R}^2 \to \mathbb{R}^1$. In this case the theorem says that most horizontal planes $\mathbb{R}^2 \times z \subset \mathbb{R}^2 \times \mathbb{R}$ are transverse to the graph of f *iff f is* C^2 (Figure 3–2). This seems plausible. In fact it seems plausible that this should hold even if f is merely C^1; but Whitney [1] has found a counterexample! In fact Whitney constructs a C^1 map $f: \mathbb{R}^2 \to \mathbb{R}^1$ whose critical set contains a topological arc I, yet $f|I$ is not constant, so that $f(\sum_f)$ contains an open subset of $\mathbb{R}$. This leads to the following paradox: The graph of f is a surface $S \subset \mathbb{R}^3$ on which there is an arc A, at every point of which the surface has a horizontal tangent plane, yet A is not at a constant height. To

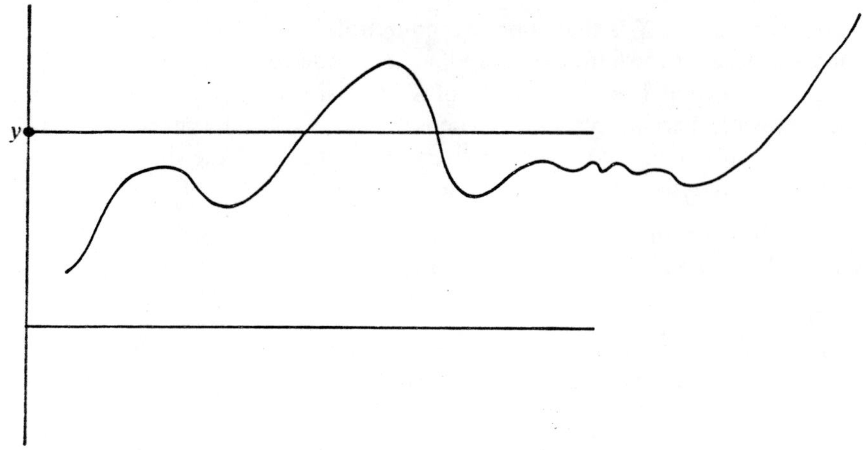

Figure 3–1.

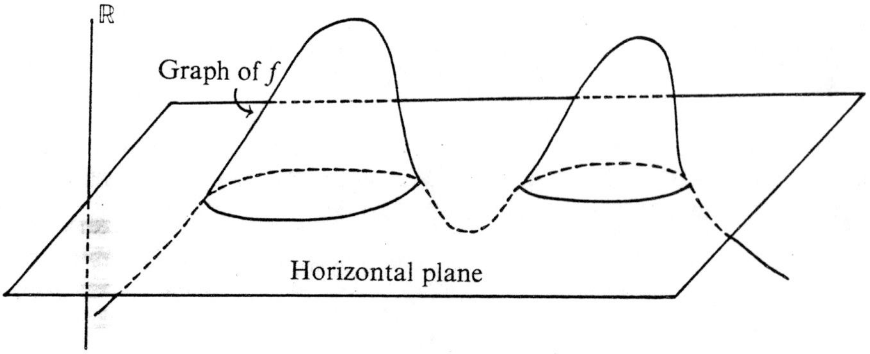

Figure 3–2.

make this more vivid, imagine that S is a hill and A a path on the hill. The hill is level at every point of the path, yet the path goes up and down.

Proof of the Morse–Sard Theorem for C^∞ Maps. It suffices to prove a local theorem; thus we deal with a C^∞ map $f:W \to \mathbb{R}^n$ where $W \subset \mathbb{R}^m$ is open. If $m < n$ then $f(W)$ has measure zero. We assume from now on that $m \geqslant n$.

A *differential operator of order* 1 means a map $C^\infty(W,\mathbb{R}) \to C^\infty(W,\mathbb{R})$ of the form

$$g \mapsto \frac{\partial g}{\partial x_k}$$

for some $k \in \{1, \ldots, m\}$. The composition of v such operators is a *differential operator of order v.*

We express the critical set $\sum_f$ as the union of three subsets as follows. Write $f(x) = (f_1(x), \ldots, f_m(x))$.

$\sum^1$ is the set of points $p \in \sum_f$ such that $\Delta f_i(p) = 0$ for all differential operators Δ of order $\leqslant m/n$ and all $i = 1, \ldots, m$.

$\sum^2$ is the set of points $p \in \sum_f$ such that $\Delta f_i(p) \neq 0$ for some i and some differential operator Δ of order $\geqslant 2$;

$\sum^3$ is the set of points $p \in \sum_f$ such that $\dfrac{\partial f_i}{\partial x_j}(p) \neq 0$ for some i, j.

Clearly $\sum_f = \sum^1 \cup \sum^2 \cup \sum^3$.

We now show that $f(\sum^1)$ has measure zero. Let v be the smallest integer such that $v > m/n$. The Taylor expansion of f of order v about points of $\sum^1$ shows that every point of $\sum^1$ has a neighborhood U in W such that if $p \in \sum^1 \cap U$ and $q \in U$ then

$$|f(p) - f(q)| \leqslant B|x - y|^v, B \geqslant 0.$$

We take U to be a cube. It suffices to prove that $f(U \cap \sum)$ has measure zero.

Let λ be the edge of U and s a large integer. Divide U into s^m cubes of edge $\lambda|s$. Of these, denote those that meet $\sum^1$ by C_k, $k = 1, \ldots, t$, where $t \leqslant s^m$.

Each C_k is contained in a ball of radius $(\lambda/s)\sqrt{m}$ centered at a point of $U \cap X$. Therefore $f(C_k)$ is contained in a cube $C'_k \subset \mathbb{R}^n$ whose edge is not more than

$$2B\left(\frac{\lambda}{s}\sqrt{m}\right)^v = A\left(\frac{\lambda}{s}\right)^v.$$

Hence the sum $\sigma(s)$ of the n-measures of these cubes C'_k is not more than

$$s^m A^n \left(\frac{\lambda}{s}\right)^{vn} = s^{m-vn} A^n \lambda^{vn}.$$

Since $m - vn < 0$ it follows that $\sigma(s) \to 0$ as $s \to \infty$. Thus $f(U \in \sum^1)$ has measure zero.

Note that $\sum^1 = \sum_f$ if $n = m = 1$. Therefore the Morse–Sard theorem is proved for this case. We proceed by induction on m. Thus we take $m > 1$ and we assume the truth of the theorem for any C^∞ map $P \to Q$ where $\dim P < m$.

We prove next that $f(\sum^2 - \sum^3)$ has measure zero. For each $p \in \sum^2 - \sum^3$ there is a differential operator θ such that

$$(1) \qquad \begin{cases} \theta f_i(p) = 0, \\ \dfrac{\partial}{\partial x_j} \theta f_i(p) \neq 0 \end{cases}$$

for some i, j. Let X be the set of such points, for fixed θ, i, j. It suffices to prove that $f(X)$ has measure zero.

Formula (1) shows that $0 \in \mathbb{R}$ is a regular value for the map $\theta f_i : W \to \mathbb{R}$. Because f_i is C^∞, so if θf_i. Therefore X is a C^∞ submanifold of dimension $m - 1$. Clearly $\sum_f \cap X \subset \sum_{f|X}$. By the induction hypothesis, $f(\sum_{f|X})$ has measure zero. Hence $f(\sum^2 - \sum^3)$ has measure zero.

It remains to prove that $f(\sum^3)$ has measure zero. Every point $p \in \sum^3$ has an open neighborhood $U \subset W$ on which, for some i, j, $\partial f_i/\partial x_j \neq 0$. By the implicit function theorem we may choose U so that there is an open set $A \times B \subset \mathbb{R}^{m-1} \times \mathbb{R}$ and a C^∞ diffeomorphism $h: A \times B \to U$ such that the following diagram commutes:

$$\begin{array}{ccc} A \times B & \xrightarrow{h} & U \\ \downarrow & & \downarrow f_i \\ B & \subset & \mathbb{R} \end{array}$$

In other words $f_i(x_1, \ldots, x_{m-1}, t) \equiv t$ for $(x,t) \in A \times B$.

For notational convenience we reorder the coordinates in $\mathbb{R}^n$ so that $f_i = f_n$. We identify U with $A \times B$ via h; now $f|U$ has the form

$$\mathbb{R}^{m-1} \times \mathbb{R} \supset A \times B \xrightarrow{f} \mathbb{R}^{n-1} \times \mathbb{R},$$
$$f(x,t) = (u_t(x),t)$$

where for each $t \in B$, $u_t: A \to \mathbb{R}^{n-1}$ is a C^∞ map. It is easy to see that (x,t) is critical for f if and only if x is critical for u_t. Thus

$$\textstyle\sum_f \cap (A \times B) = \bigcup_{t \in B} \sum_{u_t} \times t.$$

Since dim $A = n - 1$, the inductive hypothesis implies that

$$\mu_{n-1}(u_t(\textstyle\sum_{u_t})) = 0$$

where μ_{n-1} denotes Lebesgue measure in $\mathbb{R}^{n-1}$. Fubini's theorem now implies that

$$\mu_n\left(\bigcup_{t \in B} f(\textstyle\sum_{u_t} \times t)\right) = \displaystyle\int_B \mu_{n-1}(u_t(\textstyle\sum_{u_t}))\, dt$$
$$= \int_B 0\, dt = 0.$$

Therefore, reverting to the original notation, we see that $f(\sum_3 \cap U)$ has measure 0.

QED

As the first application of the Morse–Sard Theorem we prove the following topological result, equivalent to Brouwer's fixed-point theorem:

1.4. Theorem. *There is no retraction $D^n \to S^{n-1}$.*

Proof. Suppose $f: D^n \to S^{n-1}$ is a retraction, i.e., a continuous map such that $f|S^{n-1} =$ identity. We can find a new retraction $g: D^n \to S^{n-1}$ which is C^∞ on a neighborhood of S^{n-1} in D^n, for example

$$g(x) = \begin{cases} f(x/|x|) & \text{if} \quad 1/2 \leqslant |x| \leqslant 1 \\ f(2x) & \text{if} \quad 0 \leqslant |x| \leqslant 1/2. \end{cases}$$

Approximate g by a C^∞ map $h: D^n \to S^{n-1}$ which agrees with g on a neighborhood of S^{n-1}; then h is a C^∞ retraction.

By Theorem 1.4 there is a regular value $y \in S^{n-1}$ of h. (All we need is *one* regular value!) Then $h^{-1}(y)$ is a compact one-dimensional submanifold $V \subset D^n$, and

$$\partial V = V \cap S^{n-1}.$$

Therefore y is a boundary point of V. The component of V which contains y is diffeomorphic to a closed interval; *it must have another boundary point* $z \in S^{n-1}$, $z \neq y$. But $h(z) = z$, contradicting $z \in h^{-1}(y)$.

QED

The same argument proves that if M is any compact smooth manifold, there is no retraction $M \to \partial M$. This is true even without smoothness, but the proof requires algebraic topology.

Brouwer's fixed-point theorem says that any continuous map $f: D^n \to D^n$ has a fixed point, that is, $f(x) = x$ for some x. This follows from 1.5; for if $f(x) \neq x$ for all x, a retraction $g: D^n \to S^{n-1}$ is obtained by sending x to the intersection of S^{n-1} with the ray through x emanating from $f(x)$; see Figure 3–3.

This proof that D^n does not retract to S^{n-1} illustrates the interplay between maps and manifolds. The final step of the proof is the observation that a compact 1-dimensional manifold has an even number of boundary components. Thus the very simple topology of 1-manifolds leads to a highly nontrivial result about maps.

This method of studying maps is used frequently in differential topology. Its basic pattern is: approximate by a C^∞ map, find a regular value, and then exploit the topology of the inverse image of the regular value.

An important extension of the method uses not a regular value but a submanifold to which the map is transverse. To achieve transversality, further approximation theorems are needed. These are developed in the next section.

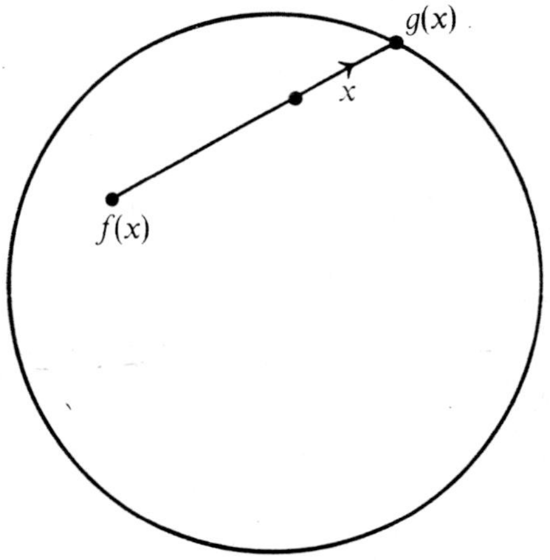

Figure 3–3.

Exercises

*1. Let $f:\mathbb{R} \to \mathbb{R}$ be differentiable (not necessarily C^1). Then the set of critical values has measure 0.

2. (a) For every $y_0 \in N$, the set

$$\{f \in C^r(M,N): y_0 \text{ is a regular value}\}$$

is open and dense in $C_S^r(M,N)$, $1 \leqslant r \leqslant \infty$.

(b) Given $y_0 \in N$, $f_0: M \to N$, a neighborhood $\mathcal{N} \subset C_S^r(M,N)$ of f_0, and a neighborhood $W \subset M$ of $f_0^{-1}(y_0)$, there exists $g \in \mathcal{N}$ such that $g \pitchfork y_0$ and $g = f_0$ on $M - W$.

3. Let M be a manifold without boundary and $K \subset M$ a closed set. Every neighborhood $U \subset M$ of K contains a closed neighborhood of K which is a smooth submanifold of M. [Consider $\lambda^{-1}[0, y]$ where $\lambda: M \to [0,1]$ is 1 on K, has support in U, and y is a regular value of λ.]

4. (a) Let M be a connected manifold and $f: M \to N$ an analytic map. Let $\sum \subset M$ be the set of critical points. If $\sum \neq M$ then $f^{-1}f(\sum)$ has measure zero.

(b) If f is merely C^∞, the conclusion of (a) can be false.

***5. Let $U \subset \mathbb{R}^3$ and $V \subset \mathbb{R}^2$ be open sets. If $f: U \to V$ is C^1 and surjective, does f necessarily have rank 2 at some point of U?

2. Transversality

Let $f: M \to N$ be a C^1 map and $A \subset N$ a submanifold. If $K \subset M$ we write $f \pitchfork_K A$ to mean that f is *transverse to A* along K, that is, whenever $x \in K$ and $f(x) = y \in A$, the tangent space N_y is spanned by A_y and the image $T_x f(M_x)$. When $K = M$ we simply write $f \pitchfork A$.

In Sections 1.3 and 1.4 it was shown that if $f \pitchfork A$ then $f^{-1}(A)$ is a submanifold (under certain restrictions on boundary behavior). This is one of the main reasons for the importance of transversality.

Define

$$\pitchfork_K^r(M,N; A) = \{f \in C^r(M,N): f \pitchfork_K A\},$$

and

$$\pitchfork^r(M,N; A) = \pitchfork_M^r(M,N; A).$$

The main result of this section is the following theorem. Recall that a *residual* subset of a space X is one which contains the intersection of countably many dense open sets. (Remember that all manifolds and submanifolds are tacitly assumed to be C^∞.)

2.1. Transversality Theorem. *Let M, N be manifolds and $A \subset N$ a submanifold. Let $1 \leqslant r \leqslant \infty$. Then:*

(a) $\pitchfork^r(M,N; A)$ *is residual (and therefore dense) in $C^r(M,N)$ for both the strong and weak topologies.*

(b) *Suppose A is closed in N. If $L \subset M$ is closed [resp. compact], then $\pitchfork_L^r(M,N; A)$ is dense and open in $C_S^r(M,N)$ [resp. in $C_W^r(M,N)$].*

The proof is based on the Morse–Sard theorem for the local result, and on the same globalization that was used for openness and density of immersions in Chapter 2. Since this technique will be used more than once, we first develop it abstractly.

Let M and N be C^r manifolds, $0 \leqslant r \leqslant \infty$. By a *$C^r$ mapping class on* (M,N) we mean a function $\mathscr{X}$ of the following type. The domain of $\mathscr{X}$ is the set of triples (L,U,V) where $U \subset M$, $V \subset N$ are open, $L \subset M$ is closed, and $L \subset U$. To each triple $\mathscr{X}$ assigns a set of maps $\mathscr{X}_L(U,V) \subset C^r(U,V)$. In addition $\mathscr{X}$ must satisfy the following *localization axiom*:

Given triples (L,U,V) a map $f \in C^r(U,V)$ is in $\mathscr{X}_L(U,V)$ provided there exist triples (L_i,U_i,V_i) and maps $f_i \in \mathscr{X}_{L_i}(U_i,V_i)$ such that $L \subset \cup L_i$ and $f = f_i$ in a neighborhood of L_i, for all i.

An example to keep in mind is:

$$\mathscr{X}_L(U,V) = \pitchfork^r_L(U,V;\, V \cap A).$$

A mapping class $\mathscr{X}$ is called *rich* if there are open covers $\mathscr{U}$, $\mathscr{V}$ of M, N such that whenever open sets $U \subset M$, $V \subset N$ are subsets of elements of $\mathscr{U}$, $\mathscr{V}$ and $L \subset U$ is compact, then $\mathscr{X}_L(U,V)$ is dense and open in $C^r_W(U,V)$.

2.2. Globalization Theorem. *Let $\mathscr{X}$ be a rich C^r mapping functor on (M,N), $0 \leqslant r \leqslant \infty$. For every closed set $L \subset M$:*

(a) *$\mathscr{X}_L(M,N)$ is dense and open in $C^r_S(M,N)$:*

(b) *$\mathscr{X}_L(M,N)$ is dense and open in $C^r_W(M,N)$ if L is compact.*

Proof. Fix L and $f \in C^r(M,N)$. In what follows i runs over a countable indexing set Λ. Let $\Phi = \{\varphi_i,U_i\}$ be a locally finite atlas on M, $K_i \subset U_i$ compact sets such that $L = \cup K_i$, and $\Psi = \{\psi_i,V_i\}$ a family of charts on N such that $f(K_i) \subset U_i$. Because $\mathscr{X}$ is rich we can choose U_i and V_i so that $\mathscr{X}_{K_i}(E,V_i)$ is dense and open in $C^r_W(E,V_i)$ for every open set $E \subset U_i$ which contains K_i.

Define $\mathscr{M} \subset C^r(M,N)$ to be the set of all $g \in C^r(M,N)$ such that

$$g|U_i \in \mathscr{X}_{K_i}(U_i,V_i) \qquad \text{for all} \qquad i \in \Lambda.$$

The localization axiom implies $\mathscr{M} \subset \mathscr{X}_L(M,N)$.

Suppose $f \in \mathscr{X}_L(M,N)$. Then $f \in \mathscr{M}$ by the localization axiom. Our assumption that each $\mathscr{X}_{K_i}(U_i,V_i)$ is weakly open implies that when L is compact $\mathscr{M}$ is weakly open (since Λ is then finite), while in general $\mathscr{M}$ is strongly open (because Φ is locally finite). This proves the two openness statements in Theorem 2.2.

We now drop the assumption that $f \in \mathscr{X}_L(M,N)$ and proceed to the density part of Theorem 2.2.

For each i let $\varepsilon_i > 0$ be given. Then there is defined the strong basic neighborhood

$$\mathscr{N} = \mathscr{N}^r(f;\, \Phi,\Psi,K,\varepsilon)$$

where $K = \{K_i\}_{i \in \Lambda}$ and $\varepsilon = \{\varepsilon_i\}_{i \in \Lambda}$.

Fix $j \in \Lambda$. Let $E = U_j \cap f^{-1}(V_j)$; then $K_j \subset E$. Since $\mathscr{X}$ is rich, $\mathscr{X}_{K_j}(E,V_j)$ is dense. Let $\lambda: E \to [0,1]$ be a C^r map with compact support, such that $\lambda = 1$ near K_j.

To simplify notation, identify V_j with an open subset of a halfspace via ψ_j, so that vector operations make sense on elements of V_j.

If $g \in C^r_W(E,V_j)$ is sufficiently close to $f|E$, the following map $\Gamma(g) = h \in C^r(M,N)$ is well-defined:

$$h(x) = \begin{cases} f(x) + \lambda(x)[g(x) - f(x)] & \text{if} \quad x \in E \\ f(x) \quad \text{if} \quad x \in M - E. \end{cases}$$

Moreover, as g tends to $f|E$ in the weak topology, $h \to f$ in the *strong* topology. Since $\mathscr{X}_{K_j}(E,V_j)$ is dense, we can choose $g \in \mathscr{X}_{K_j}(E,V_j)$ so close to $f|E$ that $h \in \mathscr{N}$. Since $h = g$ near K_j it follows that $h \in \mathscr{X}_{K_j}(M,N)$.

This shows that for all $i \in \Lambda$, the set $\mathscr{X}_{K_i}(M,N)$ is dense in $C^r_S(M,N)$; and we already know it is open. It follows from Baire that $\bigcap_i \mathscr{X}_{K_i}(M,N)$ is strongly dense; since this set contains $\mathscr{X}_L(M,N)$, the latter is therefore strongly dense.

The proof that $\mathscr{X}_L(M,N)$ is weakly dense when L is compact, is similar.

QED

The proof of Theorem 2.1 will be based on the following semilocal result.

2.3. Lemma. *Let K be a compact set in a manifold U, $\mathbb{R}^a \subset \mathbb{R}^n$ a linear subspace, and $V \subset \mathbb{R}^n$ an open set. Then*

$$\pitchfork^r_K (U,V; \mathbb{R}^a \cap V)$$

is dense and open in $C^r_W(U,V)$, $1 \leqslant r \leqslant \infty$.

Proof. Since $C^r_W(U,V)$ is open in $C^r_W(U,\mathbb{R}^n)$ it suffices to take $V = \mathbb{R}^n$.

Let $\pi: \mathbb{R}^n \to \mathbb{R}^n/\mathbb{R}^a$ be the projection. If $f \in C^r(U,\mathbb{R}^n)$ and $x \in U$, then $f \pitchfork_x \mathbb{R}^a$ if and only if: either (i) $f(x) \notin \mathbb{R}^a$, or (ii) $x \in \mathbb{R}^a$ and x is a regular point for $\pi f: U \to \mathbb{R}^n/\mathbb{R}^a$.

Suppose $f \pitchfork_K \mathbb{R}^a$. Then each $y \in K$ has a compact neighborhood $K_y \subset K$ such that either (i) holds for all $x \in K_y$ or (ii) holds for all $x \in K_y$. Let such a K_y be chosen. It is easily seen that whether (i) or (ii) holds y, the set of $f \in C^r_W(U,\mathbb{R}^n)$ such that $f \pitchfork_{K_y} \mathbb{R}^a$ is open. Since K is covered by a finite set of neighborhoods $K_{y_1}, \ldots, K_{y_s}$, it follows that $\pitchfork^r_K (U,\mathbb{R}^n; \mathbb{R}^a)$ is open.

We now prove denseness. Since C^∞ maps are dense in $C^r_W(U,\mathbb{R}^n)$, it suffices to show that an arbitrary C^∞ map $g: U \to \mathbb{R}^n$ is in the closure of $\pitchfork^r_K (U,\mathbb{R}^n; E)$. Let $\{y_k\}$ be a sequence in $\mathbb{R}^n$ tending to 0, such that each $\pi(y_k)$ is a regular value of $\pi g: U \to \mathbb{R}^n/\mathbb{R}^a$. Define

$$g_k: U \to \mathbb{R}^n,$$
$$g_k(x) = g(x) - y_k.$$

Then $g_k \to g$ in $C^r_W(U,\mathbb{R}^n)$. Since $g_k \pitchfork \mathbb{R}^a$, this shows that $\pitchfork^r_K (U,\mathbb{R}^n)$ is dense in $C^r_W(U,\mathbb{R}^n)$.

QED

Proof of the Transversality Theorem 2.1. First we assume that A is a closed submanifold and prove Theorem 2.1(b). We begin with the case $\partial N = \varnothing$.

It is easy to verify that for $1 \leqslant r \leqslant \infty$ the function

$$\mathscr{X}:(L,U,V) \mapsto \pitchfork_L^r (U,V; A \cap V)$$

is a C^r mapping class on (M,N). Under the assumption that $\partial N = \varnothing$ and A is closed, $\mathscr{X}$ is rich. This follows from Lemma 2.3 by taking $\mathscr{U}$ to be any open cover of M and $\mathscr{V}$ to be an atlas of coordinate domains on N that come from submanifold charts for (N,A). It follows from Theorem 2.2 that $\pitchfork_L^r (M,N; A)$ is strongly dense and open, and weakly dense and open if L is compact.

Suppose now that A is still closed, but $\partial N \neq \varnothing$. We may assume $N \subset \mathbb{R}^q$ as a closed submanifold. Then the weak and strong topologies on $C^r(M,N)$ are those it inherits from the weak and strong topologies $C^r(M,\mathbb{R}^q)$. We have already proved that $\pitchfork^r (M,\mathbb{R}^q; A)$ is strongly open, since $A \subset \mathbb{R}^q$ is closed and $\partial \mathbb{R}^q = \varnothing$; therefore the equation

$$\pitchfork^r (M,N; A) = C^r(M,N) \cap \pitchfork^r (M,\mathbb{R}^q; A)$$

shows that $\pitchfork^r (M,N; A)$ is strongly open in $C^r(M,N)$. A similar argument shows that $\pitchfork_L^r (M,N; A)$ is strongly open, and weakly open if L is compact. For density put $N_0 = N - \partial N$, $A_0 = A - \partial N$, so that $\partial N_0 = \varnothing$ and $A_0 \subset N_0$ is a closed submanifold. Then $\pitchfork_L^r (M,N_0; A_0)$ is strongly dense in $C^r(M,N_0)$, and weakly dense when L is compact. Now $C^r(M,N_0)$ is a subset of $C^r(M,N)$ which is both strongly and weakly dense. Therefore $\pitchfork_L^r (M,N_0; A_0)$ is a subset of $\pitchfork_L^r (M,N; A)$ which is strongly dense, and weakly dense when L is compact. This proves Theorem 2.1(b) in full generality.

To prove Theorem 2.1(a) when A is not closed, let A_k be a countable family of compact coordinate disks on A. Then

$$\pitchfork^r (M,N; A) = \bigcap_{k=1}^{\infty} \pitchfork^r (M,N; A_k).$$

Since A_k is closed, $\pitchfork^r (M,N; A_k)$ is strongly dense and open which proves $\pitchfork^r (M,N; A)$ strongly residual. Write $M = \bigcup_{j=1}^{\infty}{}' M_j$ where each M_j is compact. Then

$$\pitchfork^r (M,N; A_k) = \bigcap_{j=1}^{\infty} \pitchfork^r M_j(M,N; A_k).$$

This makes each $\pitchfork^r (M,N; A_k)$ weakly residual; hence each $\pitchfork^r (M,N; A_k)$ is weakly residual. Finally, $\pitchfork^r (M,N; A)$ is weakly residual. The proof of Theorem 2.1 is complete.

QED

Transversality is often used to put submanifolds $A, B \subset N$ in *general position*; this means that the inclusion map $B \to N$ is transverse to A, or equivalently, that $A_x + B_x = N_x$ when $x \in A \cap B$. Note that this condition is symmetric. If A and B are in general position then $A \cap B$ is a submanifold of both A and B.

2.4. Theorem. *Let A, B be C^r submanifolds of N, $1 \leqslant r \leqslant \infty$. Every neighborhood of the inclusion $i_B : B \to N$ in $C_S^r(B,N)$ contains an embedding which is transverse to A.*

Proof. From the approximation results of Chapter 2 we may assume $r = \infty$. The theorem follows from Theorem 2.1 and the openness of embeddings.

QED

Frequently one wants a map $M \to N$ to be transverse not just to one submanifold, but to each of several submanifolds $A_0, \ldots, A_q$. If each A_i is closed, the set of such maps is open and dense; this follows from openness and density of $\pitchfork^r(M,N; A_i)$. But if the A_i are not closed we may lose openness. (But see Exercises 15 and 8.)

2.5. Theorem. *Let $A_0, \ldots, A_q$ be C^r submanifolds of N, $1 \leqslant r \leqslant \infty$. The set $\pitchfork^r(M,N; A_0, \ldots, A_q)$ of C^r maps $M \to N$ that are transverse to each A_i, $i = 0, \ldots, q$, is residual in $C_S^r(M,N)$.*

Proof. Since each set $\pitchfork^r(M,N; A_k)$ is residual, $k = 0, \ldots, q$, their intersection is residual.

QED

The following typical application of transversality is frequently used. For integers $n \geqslant k \geqslant 0$, let $V_{n,k}$ denote the *Stiefel manifold* of linear maps $\mathbb{R}^k \to \mathbb{R}^n$ of rank k. It is an open submanifold of the vector space $L(\mathbb{R}^n,\mathbb{R}^k)$. An element of $V_{n,k}$ can be thought of as a *k-frame*, that is, a k-tuple of independent vectors in $\mathbb{R}^n$: the image of the standard basis of $\mathbb{R}^k$.

2.6. Theorem. *Let M be a q-dimensional manifold and $K \subset M$ a closed set. If $q \leqslant n - k$ then every map $K \to V_{n,k}$ extends to a map $M \to V_{n,k}$.*

Proof. We assume $n > k > 0$, the other cases being trivial. By covering M with a locally finite family of coordinate disks and making successive extensions, one reduces the theorem to the case $M = D^q$, which we now consider. Let $g : D^q \to L(\mathbb{R}^k,\mathbb{R}^n) - V_{n,k} = A$. By compactness, $g^{-1}(A)$ is a closed subset of D^n which is disjoint from K. By the relative approximation theorem (Theorem 2.2.5) we assume: g is C^∞ on an open set $M_0 \subset M$ containing $g^{-1}(A)$, such that K is disjoint from the closure of M_0, and $g = f$ on K.

The subset $A \subset L(\mathbb{R}^n,\mathbb{R}^k)$ is the union of the subsets

$$L(k,n; \rho) = \{T \in L(\mathbb{R}^k,\mathbb{R}^n) : \operatorname{rank} T = \rho\},$$

$\rho = 0, \ldots, k - 1$. Each of these is a submanifold. To see this fix $T \in L(k,n; \rho)$. Let $i: \mathbb{R}^\rho \to \mathbb{R}^k$ be a linear injection transverse to the kernel of T, and let $p: \mathbb{R}^n \to \mathbb{R}^\rho$ a linear surjection whose kernel is transverse to the image of T. For any $S \in L(k,n; \rho)$ in a sufficiently small neighborhood U of T, the linear map

$$pSi: \mathbb{R}^\rho \to \mathbb{R}^\rho$$

is an isomorphism. Define

$$\varphi: U \to G_{k,\,k-\rho} \times L(\mathbb{R}^\rho, \mathbb{R}^\rho) \times G_{n,\,\rho},$$
$$S \mapsto (\operatorname{Ker} S,\, pSi,\, \operatorname{im} S);$$

here $G_{m,\ell}$ is the Grassmann manifold of ℓ-planes in $\mathbb{R}^m$; its dimension is $\ell(m - \ell)$. Then φ maps U homeomorphically onto an open set. All possible maps of this type form an atlas on $L(k,n; \rho)$. Since the inverse of φ is a C^ω map into $L(k,n)$, it follows that $L(k,n; \rho)$ is a C^ω submanifold of $L(k,n)$ of dimension

$$\begin{aligned} d_\rho &= (k - \rho)\rho + \rho^2 + \rho(n - \rho) \\ &= nk - (n - \rho)(k - \rho), \end{aligned}$$

and codimension $(n - \rho)(k - \rho)$. This holds for $0 \leqslant \rho \leqslant \min\{k,n\}$. Note that d_ρ increases with ρ.

Put $A_\rho = L(k,n; \rho)$. The map g can be assumed to be transverse to each A_ρ. Now

$$d_{k-1} = (k - 1)(n + 1).$$

Therefore if $\rho \leqslant k - 1$,

$$\dim L(\mathbb{R}^k, \mathbb{R}^n) - d_\rho \geqslant kn - (k - 1)(n + 1) = n - k + 1.$$

It follows that if $\dim M < n - k + 1$, the image of g misses $A_0 \cup \cdots \cup A_{k-1}$; and so $g(M) \subset V_{n,k}$.

QED

Frequently one deals not with all C^r maps $M \to N$ but only with a family of maps parametrized by another manifold V. Thus one has a map $F: V \to C^r(M,N)$ and a submanifold $A \subset N$; it required to find $v \in V$ such that the map

$$F(v) = F_v: M \to N$$

is transverse to A. Of course restrictions must be placed on F. We call the following result the *parametric transversality theorem*. For simplicity we state it only for manifolds without boundary.

2.7. Theorem. *Let V, M, N be C^r manifolds without boundary and $A \subset N$ a C^r submanifold. Let $F: V \to C^r(M,N)$ satisfy the following conditions:*

(a) *the evaluation map $F^{\mathrm{ev}}: V \times M \to N$, $(v,x) \mapsto F_v(x)$, is C^r;*

(b) *F^{ev} is transverse to A;*

(c) $r > \max\{0, \dim N + \dim A - \dim M\}$.

Then the set

$$\pitchfork(F; A) = \{v \in V : F_v \pitchfork A\}$$

is residual and therefore dense. If A is closed in N and F is continuous for the strong topology on $C^r(M,N)$ then $\pitchfork(F,A)$ is also open.

Proof. The last statement follows from openness of $\pitchfork^r(M,N; A)$. To prove the rest of the theorem let $W = (F^{ev})^{-1}(A) \subset V \times M$. By (a), W is a C^r submanifold of $V \times M$. Let $\pi: V \times M \to V$ be the projection. It is readily verified that $F_v \pitchfork A$ is and only if $v \in V$ is a regular value for the C^r map $\pi|W: W \to V$. The dimension of W is dim V + dim M − dim N + dim A. The theorem follows from Morse–Sard.

QED

In many situations parametric transversality is not sufficient; instead of a map from a manifold to a function space, one must deal with a map defined on another function space. Often the domain of the map is an infinite dimensional manifold; in this case there is a generalization of Theorem 2.3 due to R. Abraham [1].

A common situation concerns the jet map

$$j^r : C^s(M,N) \to C^{s-r}(M,J^r(M,N)).$$

Here $1 \leqslant r < s \leqslant \infty$. One is given a submanifold $A \subset J^r(M,N)$ and tries to approximate a C^s map $g: M \to N$ by another C^s map h whose prolongation $j^r h: M \to J^r(M,N)$ is transverse to A. Denote the set of such maps h by $\pitchfork^s(M,N; j^r, A)$.

2.8. Jet Transversality Theorem. *Let M, N be C^∞ manifolds without boundary, and let $A \subset J^r(M,N)$ be a C^∞ submanifold. Suppose $1 \leqslant r < s \leqslant \infty$. Then $\pitchfork^s(M,N; j^r, A)$ is residual and thus dense in $C^s_S(M,N)$, and open if A is closed.*

Proof. Suppose A is closed. Openness follows from openness of $\pitchfork^{s-r}(M,J^r(M,N); A)$. To prove density let $U \subset M$, $V \subset N$ be open sets and let $L \subset U$ be closed in M. Define

$$\pitchfork_L(U,V) = \{f \in C^s(U,V) : j^r f \pitchfork_L A\}.$$

One verifies easily that $\mathscr{X}$ is a C^s mapping class on (M,N). By the globalization Theorem 2.2 it suffices to prove $\mathscr{X}$ rich. For the coverings $\mathscr{U}$, $\mathscr{V}$ in the definition of rich choose any open coverings by coordinate domains. It now suffices to prove that if $U \subset \mathbb{R}^m$ is an open subset and $A \subset J^r(U,\mathbb{R}^n)$ is a closed submanifold, then $\pitchfork^s(U,\mathbb{R}^n; j^r, A)$ is open and dense in $C^s_W(U,\mathbb{R}^n)$. Moreover it is enough to prove this for s finite, $s > r$. Fix $f \in C^s(U,\mathbb{R}^n)$. Openness is obvious. The strategy for denseness is to find a C^∞ manifold X and a map $\alpha: X \to C^s_W(U,\mathbb{R}^n)$ with $f \in \alpha(x)$, and then apply parametric transversality to the composition

$$F: X \xrightarrow{\alpha} C^s(U,\mathbb{R}^n) \xrightarrow{j^r} C^{s-r}(U,J^r(U,\mathbb{R}^n)).$$

This requires that the evaluation map of F,

$$F^{ev}: X \times U \to J^r(U,\mathbb{R}^n)$$

be transverse to A and sufficiently differentiable. In fact F^{ev} will be a C^∞ submersion.

Put $X = J^s_0(\mathbb{R}^m,\mathbb{R}^n)$. Every element of X is the s-jet at 0 of a unique map $g:\mathbb{R}^m \to \mathbb{R}^n$ whose coordinate maps $g_1, \ldots, g_n$ are polynomials of degree $\leqslant s$ in the coordinates of $\mathbb{R}^m$. We identify the elements of X with such maps g.

Define $\alpha: X \to C^s_W(U,\mathbb{R}^n)$, $g \mapsto f + g|U$ and

$$F = j^r \circ \alpha: X \to C^{s-r}(U, J^r(U,\mathbb{R}^n)).$$

Then $F(0) = f$. To compute F^{ev} make the natural identification

$$J^r(U,\mathbb{R}^n) = J^r_0(\mathbb{R}^m,\mathbb{R}^n) \times U.$$

Then

$$F^{ev}: J^s_0(\mathbb{R}^m,\mathbb{R}^n) \times U \to J^r_0(\mathbb{R}^m,\mathbb{R}^n) \times U$$

is given by

$$(j^s_0(g),x) \mapsto (j^r_0(g + f),x).$$

The map

$$\beta: J^s_0(\mathbb{R}^m,\mathbb{R}^n) \to J^r_0(\mathbb{R}^m,\mathbb{R}^n)$$
$$j^s_0(g) \mapsto j^r_0(g + f)$$

is affine, hence F is C^∞. Moreover the derivative of β at any point is the "forgetful" linear map

$$J^s_0(\mathbb{R}^m,\mathbb{R}^n) \to J^r_0(\mathbb{R}^m,\mathbb{R}^n),$$
$$j^s_0(g) \mapsto j^r_0(g)$$

which is surjective. Thus $F^{ev} \pitchfork A$; by Theorem 2.7 it follows that

$$\{x \in X : j^r(\alpha(x)) \pitchfork A\}$$

is dense in X. Since

$$\alpha: X \to C^s_W(U,\mathbb{R}^n)$$

is continuous it follows that f is in the closure of

$$\{h \in C^s_W(U,\mathbb{R}^n) : j^r h \pitchfork A\}.$$

This proves that $\mathscr{X}$ is rich; hence for closed A, Theorem 2.8 follows from Theorem 2.2. If A is not closed write $A = \bigcup_{k=1}^{\infty} A_k$ where each A_k is a compact coordinate disk in A. Then each $\pitchfork^s(M,N; j^r, A_k)$ is dense and open in $C^s_S(M,N)$. By Baire their intersection, which is $\pitchfork^s(M,N; j^r, A)$, is dense.

QED

Just as with ordinary transversality, jet transversality extends to submanifold families; we leave the proof of the following result to the reader.

2.9. Theorem. *Let $A_0, \ldots, A_q$ be C^∞ submanifolds of $J^r(M,N)$. If $1 \leqslant r < s \leqslant \infty$ then the set*

$$\{f \in C^s(M,N): j^r f \pitchfork A_k, k = 0, \ldots, q\}$$

is residual in $C_S^s(M,N)$.

As an application consider anew the question of density of immersions in $C_S^2(M,N)$. Let $A_k \subset J^1(M,N)$ be the set of 1-jets of rank k. Let $m = \dim M$, $n = \dim N$. Then $A_0, \ldots, A_{m-1}$ is a C^∞ submanifold family. A map $f: M \to N$ is an immersion if and only if the image of $j^1 f$ misses $A_0 \cup \cdots \cup A_{m-1}$. The set of $f \in C^2(M,N)$ such that $j^1 f$ is transverse to $A_0, \ldots, A_{m-1}$ is dense in $C_S^2(M,N)$. If, for $i = 0, \ldots, m-1$, $\dim A_i + \dim M < \dim J^1(M,N)$, such transversality implies that f is an immersion. As in the proof of Theorem 2.5 one computes that (assuming $m \leqslant n$)

$$\dim A_i \leqslant \dim A_{m-1} = 2m + mn - 1,$$
$$\dim J^1(M,N) = mn + m + n.$$

Density of immersions in thus implied by

$$(2m + mn - 1) + m < mn + m + n$$

which is the same as $n \geqslant 2m$, exactly the condition found previously.

This proof is not very satisfying geometrically; it gives no hint as to how the immersion is constructed. Nevertheless it shows the power of transversality: the existence and even the denseness of immersions is proved by merely counting dimensions!

Exercises

1. An immersion $f: M \to N$ has *clean double points* if whenever x, y are distinct points of M with $f(x) = f(y)$, they have disjoint neighborhoods U, V such that $f|U$ and $f|V$ are embeddings, and the submanifolds $f(U)$, $f(V)$ are in general position (as defined in Section 3.2). The set of immersions that have clean double points is dense and open in $\mathrm{Imm}_S^r(M,N)$, $1 \leqslant r \leqslant \infty$.

2. An immersion $f: M \to N$ *is in general position* if for any integer $k \geqslant 2$, whenever $f(x_1) = \cdots = f(x_k) = y$ and the points $x_1, \ldots, x_k$ are distinct, then N_y is spanned by $Tf(M_{x,k})$ and

$$Tf(M_{x_1}) \cap \cdots \cap Tf(M_{x_{k-1}}).$$

The set of proper immersions which are in general position is dense and open in $\mathrm{Imm}_S^r(M,N)$, $1 \leqslant r \leqslant \infty$.

3. If $f: M \to N$ is transverse to a submanifold complex $A_0, \ldots, A_q$ then $f^{-1}(A_0 \cup \cdots \cup A_q)$ is a submanifold complex (see Ex. 15).

4. There is a dense open set $\mathscr{S} \subset C_S^\infty(M^m,N^n)$ such that if $f \in \mathscr{S}$, then:
(a) for each $\rho = 0, \ldots, \min(m,n)$ the set

$$R(f,\rho) = \{x \in M^m: \operatorname{rank} T_x f = \rho\}$$

is a submanifold of M;
(b) $R(f,\rho) = \varnothing$ if $(m - \rho)(n - \rho) > m$;

(c) if $(m - \rho)(n - \rho) \leqslant m$ then

$$\operatorname{codim} R(f,\rho) = (m - \rho)(n - \rho);$$

(d) the submanifolds

$$R(f,0), \ldots, R(f,\min(m,n))$$

form a submanifold complex (see Ex. 15).

5. Generically, a C^1 map $f:M^m \to N^{2m-1}$ has rank $> m - 2$ everywhere, and the set where f has rank $m - 1$ is a closed 0-dimensional submanifold (perhaps empty).

***6.** A map $f:\mathbb{R}^2 \to \mathbb{R}^2$ has a *cusp* at $x \in \mathbb{R}^2$ if (i) Df_x has rank 1, (ii) $j^1 f$ is transverse at x to the 1-jets of rank 1, and (iii) Ker Df_x is tangent to $R(f,1)$ (see Exercise 4).

(a) (0,0) is a cusp of the map $g(x,y) = (x^3 - xy, y)$.

(b) If $U \subset \mathbb{R}^2$ is any neighborhood of (0,0), there is a weak C^2 neighborhood $\mathcal{N}$ of g such that every map in $\mathcal{N}$ has a cusp in U.

***7.** A *k-fold point* of a map $f:M \to N$ is the set of $x \in M$ such that there are k distinct points $x = x_1, \ldots, x_k$ with $f(x_1) = \cdots = f(x_k)$. Let M and N be manifolds such that

$$\frac{k+1}{k} < \frac{\dim N}{\dim M} \leqslant \frac{k}{k-1}, \qquad k \geqslant 2.$$

(a) There is a dense open set of maps in $C_S^r(M,N)$, $1 \leqslant r \leqslant \infty$, having no $(k+1)$-fold points, and whose set of k-fold points is a closed C^r submanifold of dimension $km - (k - 1)n$ (possibly empty).

(b) There is a nonempty open set of maps in $C_S^r(M,N)$, each having a nonempty set of k-fold points.

8. The transversality Theorems 2.5, 2.8, 2.9, combined with Ex. 15, take the following forms for weak topologies:

(a) In Theorem 2.5 the set of maps $M \to N$ transverse to $A_0, \ldots, A_q$ is residual in $C_W^r(M,N)$, and open if $\cup A_i$ is compact and $\{A_i\}$ is a submanifold complex.

(b) In Theorem 2.8, $\pitchfork^s(M,N; j^r, A)$ is residual in $C_W^s(M,N)$, and open if A is compact.

(c) In Theorem 2.9, the set of maps whose r-jets are transverse to $A_0, \ldots, A_q$ is residual in $C_W^s(M, N)$, and open if $\cup A_i$ is compact and $\{A_i\}$ is a submonifold complex.

9. Consider $G_{s,k}$ embedded in $G_{s+1,k+1}$ by identifying a k-plane $P \subset \mathbb{R}^s$ with $P \times \mathbb{R} \subset \mathbb{R}^s \times \mathbb{R} = \mathbb{R}^{s+1}$. If $\dim M \leqslant k$, every map $f:M \to G_{s+1,k+1}$ is homotopic to a map $g:M \to G_{s,k}$. If $\dim M < k$, the homotopy class of g is uniquely determined by that of f.

10. Let $F:V \to C^r(M,N)$ be such that $F^{\text{ev}}:V \times M \to N$ is C^r (see the parametric transversality Theorem 2.7). Then F is continuous for the strong topology if V is compact; or, more precisely, if and only if F is constant outside a compact subset of V.

11. In the jet transversality Theorem 2.8, the assumption that $A \subset J^r(M,N)$ be a C^∞ submanifold can be relaxed to: A is a C^k submanifold, for a certain $k < \infty$ depending on r, s, $\dim M$, and $\dim N$. Compute k.

*****12.** Are the parametric and jet transversality Theorems 2.7 and 2.8 true when V, M, N, and A are allowed to have boundaries? (The proof of Theorem 2.8 uses Theorem 2.7. In Theorem 2.7 there are two difficulties: the first is that $V \times M$ is not a manifold if V and M are ∂-manifolds; the second, and more troublesome, is that $(F^{\text{ev}})^{-1}(A)$ might not be a submanifold if N and A are ∂-manifolds.)

13. Let $p:V \to M$ be a C^1 submersion, and $f:M \to V$ a C^r section of p (that is, $pf = 1_M$), $1 \leqslant r \leqslant \infty$. Let $A \subset V$ be a C^r submanifold. Then every neighborhood of f in $C_S^r(M,V)$ contains a C^∞ section transverse to A.

14. Let $g: A \to N$ be a map. A map $f: M \to N$ is *transverse to g*, written $f \pitchfork g$ if, whenever $f(x) = g(y) = z$, the images of $T_x f$ and $T_y g$ span N_z.

(a) $f \pitchfork g$ if and only if the map $f \times g: M \times A \to N \times N$ is transverse to the diagonal.

***(b) Is it true (as seems likely) that the set $\{f \in C^r(M,N): f \pitchfork g\}$ is residual in $C_S^r(M,N)$ and open if g is proper?

15. Submanifolds $A_0, \ldots, A_q \subset N$ form a *submanifold complex* if $(i) A_0$ is closed and

$$\bar{A}_{i+1} - A_{i+1} \subset A_0 \cup \cdots \cup A_i;$$

(ii) $\dim A_{i-1} < \dim A_i$; (iii) Let $0 \leqslant i < j \leqslant q$; put $d = \dim A_i$. If a sequence $\{x_n\}$ in A_j converges to y in A_i, there is a sequence E_n of d-planes, $E_n \subset T_{x_n} A_j$, converging to $T_y A_i$.

(a) The set of C^r maps $M \to N$ transverse to all the A_i, is dense and open.

*(b) The submanifolds A_ρ in the proof of Theorem 2.5 form a submanifold complex.

Chapter 4

Vector Bundles and Tubular Neighborhoods

The paradox is now fully established that the utmost abstractions are the true weapons with which to control our thought of concrete fact.

—A. N. Whitehead, *Science and the Modern World*, 1925

The Committee which was set up in Rome for the unification of vector notation did not have the slightest success, which was only to have been expected.

—F. Klein, *Elementary Mathematics from an Advanced Standpoint*, 1925

Deux surfaces fermées, par example de genre 0, situées dans une variété à 4 dimensions, sont toujours equivalentes, mais, comme nous le voyons, leurs *entourages* ne le sont pas nécessairement.

—Heegard, Dissertation, 1892

Although the concept of tangent bundle was defined in the first chapter, until now we have made only minimal use of it. In this chapter we abstract certain features of the tangent bundle, thus defining a mixed topological–algebraic object called a vector bundle. Most of the deep invariants of a manifold are intimately linked to the tangent bundle; their development requires a general theory of vector bundles.

A vector bundle can be thought of a family $\{E_x\}_{x \in B}$ of disjoint vector spaces parameterized by a space B. The union of these vector spaces is a space E, and the map $p: E \to B$, $p(E_x) = x$ is continuous. Moreover p is *locally trivial* in the sense that locally (with respect to B), E looks like a product with $\mathbb{R}^n$: there are open sets U covering B and homeomorphisms $p^{-1}(U) \approx U \times \mathbb{R}^n$, mapping each fibre E_x linearly onto $x \times \mathbb{R}^n$. A *morphism* from one vector bundle to another is a map taking fibres linearly into fibres.

A vector bundle is similar to a manifold in that both are built up from elementary objects glued together by maps of a specified kind. For manifolds the elementary objects are open subsets of $\mathbb{R}^n$; the gluing maps are diffeomorphisms. For vector bundles the elementary objects are "trivial" bundles $U \times \mathbb{R}^n$; the gluing maps are morphisms $U \times \mathbb{R}^n \to U \times \mathbb{R}^n$ of the form $(x,y) \mapsto (x,g(x)y)$ where $g: U \to GL(n)$.

In Section 4.1 the basic definitions are given and the covering homotopy theorem is proved. This basic result is the link between vector bundles and homotopy.

In both manifolds and vector bundles, linear maps play a crucial role. But whereas linear maps enter into manifolds in a rather subtle way, as derivatives, the linearity in vector bundles is closer to the surface. This makes the category of vector bundles far more flexible than that of manifolds; as a consequence, vector bundles are considerably easier to analyze. Many natural constructions can be made with vector bundles which are impossible for manifolds, such as direct sum, quotients and pullbacks. These are discussed in Section 4.2.

In Section 4.3 we prove an important classification theorem for vector bundles. This theorem says that for given integers k; $n \geqslant 0$ there is an explicitly defined k-plane bundle $\xi \to G$ which is *universal* in the following sense: for every k-plane bundle $\eta \to M$ where M is a manifold of dimension $\leqslant n$, there is a map $f: M \to G$ such that ξ is isomorphic to $f^*\xi$ (the pullback of η by f), and f is unique up to homotopy. This means that isomorphism classes of k-plane bundles over M are in natural one-to-one correspondence with homotopy classes of maps $M \to G$. In this way all questions about vector bundles over M are translated into questions about homotopy classes of maps $M \to G$.

Section 4.4 introduces the important concept of orientation for vector spaces, vector bundles and manifolds. The orientability or nonorientability of a manifold is an important invariant. As applications some nonembedding theorems are proved.

In Sections 4.5 and 4.6 a new connection between vector bundles and the topology of manifolds is introduced: the tubular neighborhood. If $M \subset N$ is a neat submanifold, M has a neighborhood in N which looks like the normal vector bundle of M in N; moreover, such neighborhoods are essentially unique. Thus the study of the kinds of neighborhoods that M can have as a submanifold of a larger manifold, is reduced to the classification of vector bundles over M. For example, the problem of whether the inclusion $M \hookrightarrow N$ can be approximated by embeddings $M \hookrightarrow N - M$ is equivalent to the problem of whether the normal bundle of M in N has a nonvanishing section.

Section 4.7 exploits tubular neighborhoods to prove that every compact manifold without boundary has a compatible real analytic structure.

1. Vector Bundles

Let $p: E \to B$ be a continuous map. A *vector bundle chart* on (p,E,B) with *domain* U and *dimension* n a homeomorphism $\varphi: p^{-1}(U) \approx U \times \mathbb{R}^n$ where $U \subset B$ is open, such that the diagram

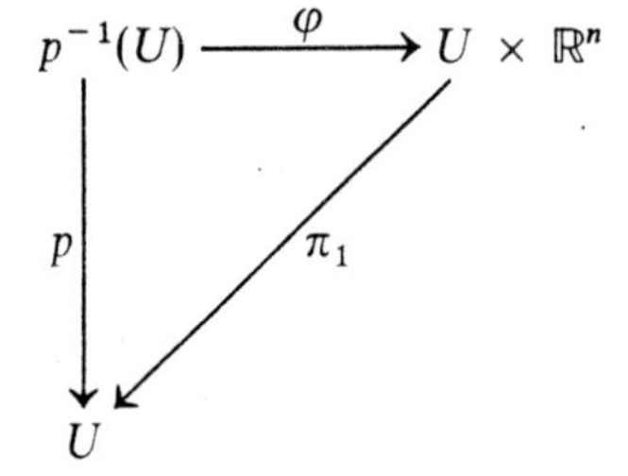

commutes; here $\pi_1(x,y) = x$. For each $x \in U$ we define the homeomorphism φ_x to be the composition

$$\varphi_x : p^{-1}(x) \xrightarrow{\varphi} x \times \mathbb{R}^n \to \mathbb{R}^n.$$

Thus if $y \in p^{-1}(x)$ we have the formula

$$\varphi(y) = (x, \varphi_x(y)).$$

A *vector bundle atlas* Φ on (p,E,B) is a family of vector bundle charts on (p,E,B) with values in the same $\mathbb{R}^n$, whose domains cover B, and such that whenever (φ,U) and (ψ,V) are in Φ and $x \in U \cap V$, the homeomorphism

$$\psi_x \varphi_x^{-1} : \mathbb{R}^n \to \mathbb{R}^n$$

is linear. The map

$$U \cap V \to GL(n),$$
$$x \to \psi_x \varphi_x^{-1}$$

is required to be continuous; it is called the *transition function* of the pair of charts (φ,U), (ψ,V). If $\Phi = \{\varphi_i, U_i\}_{i \in \Lambda}$ we obtain a family $\{g_{ij}\}$ of transition function,

$$g_{ij} : U_i \cap U_j \to GL(n).$$

These maps satisfy the identities

$$g_{ij}(x) g_{jk}(x) = g_{ik}(x) \qquad (x \in U_i \cap U_j \cap U_k),$$
$$g_{ii}(x) = 1 \in GL(n).$$

The family $\{g_{ij}\}$ is also called the *cocycle* of the vector bundle atlas Φ. A maximal vector bundle atlas Φ is a *vector bundle structure* on (p,E,B). We then call $\xi = (p,E,B,\Phi)$ a *vector bundle* having (*fibre*) *dimension n*, *projection p*, *total space E* and *base space B*. Often Φ is not explicitly mentioned. In fact we may denote ξ by E, or E by ξ. Sometimes it is convenient to put $E = E\xi$, $B = B\xi$, etc. An atlas for ξ will mean a subatlas of Φ.

The *fibre* over $x \in B$ is the space $p^{-1}(x) = \xi_x = E_x$. We give ξ_x the vector space structure making each $\varphi_x : \xi_x \to \mathbb{R}^n$ an isomorphism; this structure is independent of the choice of $(\varphi,U) \in \Phi$. Thus E is a "bundle" of vector spaces. To indicate the dimension n we sometimes call ξ an *n-plane bundle*.

If $A \subset B$ is any subset we may denote $p^{-1}(A)$ by ξ_A, $\xi|A$, E_A, or $E|A$. The *restriction* of ξ to A is the vector bundle

$$\xi|A = (p|E_A, E_A, A, \Phi_A)$$

where Φ_A contains all charts of the form

$$\varphi|p^{-1}(A \cap U) : E|A \cap U \to (A \cap U) \times \mathbb{R}^n,$$

where $(\varphi,U) \in \Phi$.

The *zero section* of ξ is the map $Z : B \to E$ which to x assigns the zero element of ξ_x. Often we call the subspace $Z(B) \subset E$ the zero section. It is frequently useful to identify B with $Z(B)$ via Z.

Let $\xi_i = (p_i,E_i,B_i,\Phi_i)$ be a vector bundle, $i = 0, 1$. A *fibre map* $F:\xi_0 \to \xi_1$ is a map $F:E_0 \to E_1$ which covers a map $f:B_0 \to B_1$, that is, there is a commuting diagram

$$\begin{array}{ccc} E_0 & \xrightarrow{F} & E_1 \\ {\scriptstyle p_0}\downarrow & & \downarrow{\scriptstyle p_1} \\ B_0 & \xrightarrow[f]{} & B_1 \end{array}$$

Thus if $x \in B_0$ and $f(x) = y$, then F maps the fibre over x into the fibre over y by a map $F_x:\xi_{0x} \to \xi_{1y}$.

If each map F_x is linear we call F a *morphism* of vector bundles. If F is a morphism and each F_x is injective, F is a *monomorphism*; if each F_x is surjective, F is an *epimorphism*; while if each F_x is bijective we call F a *bimorphism* or *vector bundle map*. If F is a bimorphism covering a homeomorphism $f:B_0 \to B_1$ then F is an *equivalence*. If $B_0 = B_1 = B$ and $f = 1_B$ then F is an *isomorphism*, and we may write $\xi_0 \cong \xi_1$.

The *trivial* n-dimensional vector bundle over B is

$$\varepsilon_B^n = (p,B \times \mathbb{R}^n,B,\Phi)$$

where $p:B \times \mathbb{R}^n \to B$ is the natural projection and Φ is the unique maximal vector bundle atlas containing the identity map of $B \times \mathbb{R}^n$. More generally, a vector bundle ξ is called trivial if it is isomorphic to ε_B^n. Such an isomorphism is a *trivialization* of ξ.

Fix a differentiability class C^r, $1 \leqslant r \leqslant \omega$. The above definitions make sense if all spaces involved are required to be C^r manifolds, and maps are required to be C^r maps. In this way we obtain C^r vector bundles, morphisms and bundle maps. We also interpret C^0 vector bundle morphism to mean vector bundle as originally defined; similarly for C^0 morphisms, etc. We denote C^r isomorphism by $\cong_r$.

The prime example of a C^r vector bundle is the tangent bundle $p:TM \to M$ of a C^{r+1} manifold M. For each chart $\psi:U \to \mathbb{R}^n$ we define a vector bundle chart

$$p^{-1}(U) \to U \times \mathbb{R}^n$$

by sending the tangent vector $X \in T_xM$ to $(x,D\varphi_x(X))$. If $f:M \to N$ is a C^{r+1} map then $Tf:TM \to TN$ is a C^r vector bundle morphism. Note that Tf is a monomorphism, epimorphism or equivalence according as f is an immersion, submersion or diffeomorphism.

If TM is trivial M is called *parallelizable*.

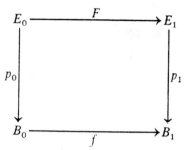

There is evidently a category of C^r vector bundles and C^r morphisms. An isomorphism in this category is an equivalence of vector bundles. For each C^r manifold M there is the subcategory of C^r vector bundles over M and C^r morphisms over 1_M (for $r = 0$, M can be any space). An isomorphism in this subcategory is an isomorphism of vector bundles. The tangent functor T is a covariant functor from the category of C^{r+1} manifolds to the category of C^r vector bundles.

The following lemma is the first step in the proof of the covering homotopy theorem.

1.1. Lemma. *Let $\xi = (p,E,B \times I)$ be a C^r vector bundle, $0 \leqslant r \leqslant \infty$. Then each $b \in B$ has a neighborhood $V \subset B$ such that $\xi|V \times I$ is trivial.*

Proof. By compactness of I and local triviality of ξ we can find a neighborhood $V_i \subset B$ of b and a subdivision of I into intervals $I_i = [t_{i-1},t_i]$, $0 = t_0 < \cdots < t_m = 1$ such that ξ is trivial over a neighborhood of $V_i \times [t_{i-1},t_{i-1}]$, $i = 1,\ldots,m$. Put $V = \cap V_i$; then I_i has a neighborhood $U_i \subset I$ such that $\xi|V \times U_i$ is trivial.

We proceed by induction on m; if $m = 1$ there is nothing more to prove. Therefore we shall show that if $m > 1$, there is a neighborhood $J \subset I$ of $[0,t_2]$ such that $\xi|V \times J$ is trivial. Continuing in this way will eventually show that $\xi|V \times I$ is trivial. Hence it suffices to assume that $m = 2$.

Let $U_1 = [0,b]$, $U_2 = [a,1]$, $0 < a < b < 1$. Choose C^r trivializations

$$\varphi_i : \xi|(V \times U_i) \to (V \times U_i) \times \mathbb{R}^n, \qquad i = 1, 2.$$

Define a C^r map

$$g: V \times [a,b] \to GL(n),$$
$$g(x) = \varphi_{1x}\varphi_{2x}^{-1}, \qquad x \in V \times [a,b].$$

Next we construct a C^r map

$$h: V \times [a,1] \to GL(n)$$

such that $h = g$ on $V \times [a,c]$ for some c, $a < c < b$. Let $\lambda:[a,1] \to [a,b]$ be a C^r map which is the identity on a neighborhood $[a,c]$ of a.

Put $\mu = 1_V \times \lambda: V \times [a,1] \to V \times [a,b]$. Define $h = g \circ \mu$.

Finally define, for each $x \in V \times I$:

$$\psi_x : \xi_x \to \mathbb{R}^n,$$

$$\psi_x = \begin{cases} \varphi_{1x} & \text{if} \quad x \in V \times [0,c] \\ h(x)\varphi_{2x}^{-1} \text{ (multiplication in } GL(n)) & \text{if} \quad x \in V \times [a,1]. \end{cases}$$

The two definitions agree for $x \in [a,c]$. Hence the maps ψ_x fit together to give a C^r trivialization of ξ.

QED

1.2. Corollary. *Every C^r vector bundle $(0 \leqslant r \leqslant \infty)$ over an interval is trivial.*

The proof of the covering homotopy theorem is based on the following C^r version of the homotopy extension property:

1.3. Lemma. *Fix $0 \leqslant r \leqslant \infty$ and let N, P be topological spaces which are C^r manifolds if $r > 0$. If $r = 0$ suppose also that N is a normal space. Let $Z \subset N$ be closed and $V \subset U \subset N$ be open, with $Z \subset V \subset \bar{V} \subset U$. Suppose given a commuting diagram*

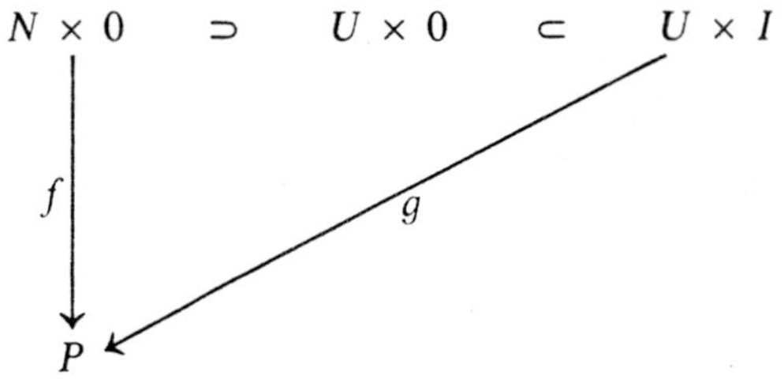

where f and g are C^r. Then there exists a C^r map $h: N \times I \to P$ such that $h|N \times 0 = f$ and $h = g$ on $V \times I$.

Proof. Let $\lambda: N \to [0,1]$ be a C^r map with support in U such that $\lambda(V) = 1$. Define a C^r map

$$h: N \times I \to P,$$

$$h(x,t) = \begin{cases} g(x,\lambda(x)t) & \text{if} \quad x \in U \\ f(x) & \text{if} \quad x \in N - U. \end{cases}$$

Then h has the required properties.

QED

The following corollary of Theorem 1.3 is called the *homotopy extension theorem.*

1.4. Theorem. *Let Z be a closed subspace of a normal space N. Let $f: N \to P$ be a continuous map and let $g: Z \times I \to P$ be a homotopy of $f|Z$. If g extends to a homotopy of $f|U$, for some neighborhood $U \subset N$ of Z, then g extends to a homotopy $h: N \times I \to P$ of f. In particular this is the case if Z is a retract of an open subset of N.*

The following theorem is the basic connection between vector bundles and homotopy. For reasons to be explained later it is called the *covering homotopy theorem.*

1.5. Theorem. *Let ξ be a C^r vector bundle over $B \times I$, $0 \leqslant r \leqslant \infty$. Assume B is paracompact. Then ξ is C^r isomorphic to the vector bundle $(\xi|B \times 0) \times I$.*

Proof. Put $\xi|B \times 0 = \eta = (p,E,B)$. Let $\eta \times I = (p \times 1_I, E \times I, B \times I)$. We shall construct a bundle map $\xi \to \eta \times I$ over the identity map of $B \times I$.

For this we use the globalization Theorem 2.2.11 applied to a suitable structure functor on B.

Let $\mathscr{X} = \{X_i\}$ be a locally finite closed covering of B by sets X_i having the following property: each X_i has a neighborhood $V_i \subset B$ such that $\xi|V_i \times I$ is trivial (use Theorem 1.2). It follows that $\eta|V_i \times I$ is also trivial. Let $\mathfrak{Y}$ be the family of unions of elements of $\mathscr{X}$.

Let $Y \subset B$. Consider pairs (f,N) where $N \subset B$ is a neighborhood of Y and $f:\xi|N \times I \to (\eta|N) \times I$ is a C^r isomorphism. Two pairs (f_i,N_i), $i = 0, 1$, *have the same Y-germ* if Y has a neighborhood $M \subset N_0 \cap N_1$ such that $f_0 = f_1$ on $\xi|M \times I$. This is an equivalence relation; an equivalence class is called a *Y-germ.* If $Y \in \mathfrak{Y}$ the set of all Y-germs is denoted by $\mathscr{F}(Y)$.

If $Z \in \mathfrak{Y}$ and $Y \subset Z$, restriction defines a map

$$\mathscr{F}_{YZ}:\mathscr{F}(Z) \to \mathscr{F}(Y).$$

In this way a structure functor $(\mathscr{F},\mathfrak{Y})$ on B is defined.

It is evident that $(\mathscr{F},\mathfrak{Y})$ is continuous; and Theorem 1.1 implies it is nontrivial. In fact Theorem 1.1 also makes $(\mathscr{F},\mathfrak{Y})$ locally extendable. For let $X \in \mathscr{X}$, $Y \in \mathfrak{Y}$. We must prove that

$$\mathscr{F}_{X,Y \cup X}:\mathscr{F}(Y \cup X) \to \mathscr{F}(X)$$

is surjective. This amounts to extending every $X \cap Y$ – germ to an X germ. Now $X \times I$ has a neighborhood $N \times I$ over which both ξ and $\eta \times I$ are trivial. Since isomorphisms of the trivial bundle are the same as maps into $GL(n)$, local extendability is implied by the following statement: if $U \subset N$ is a neighborhood of $X \cap Y$ and

$$g:(U \times I, U \times 0) \to (GL(n),1)$$

is a C^r map (where $1 \in GL(n)$ is the identity matrix), then there is a neighborhood $V \subset U$ of $X \cap Y$ and a C^r map

$$(N \times I, N \times 0) \to (GL(n),1)$$

which agrees with g on $V \times I$. But this is a consequence of Theorem 1.3; therefore $(\mathscr{F},\mathfrak{Y})$ is locally extendable.

We now apply Theorem 2.2.11 and conclude that $\mathscr{F}(B)$ is nonempty. This is equivalent to Theorem 1.4.

QED

1.6. Corollary. *Two C^r vector bundles ξ_0, ξ_1 over a paracompact base space B are C^r isomorphic if and only if there is a C^r vector bundle η over $B \times I$ such that*

$$\xi_i \cong_r \eta|B \times 0 \qquad (i = 0, 1).$$

Proof. If η exists, $\xi_0 \cong_r \xi_1$ by Theorem 1.3. Conversely, if $F:\xi_0 \to \xi_1$ is a C^r isomorphism we can take $\eta = \xi_0 \times I$.

QED

Exercises

1. Let ξ, η be vector bundles over a paracompact space and let $A \subset B$ be closed. Then every morphism $f: \xi|A \to \eta|A$ over 1_A extends to a morphism $g: \xi|W \to \eta|W$ over 1_W, for some neighborhood $W \subset B$ of A, and if f is a mono-, epi-, or bimorphism so is g.

2. Let $\xi \to B$ be a vector bundle over a paracompact space, and let $A \subset B$ a contractible closed subset. Then A has a neighborhood $W \subset B$ such that $\xi|W$ is trivial.

3. Exercises 1 and 2 are true in the category of C^r bundles, $1 \leqslant r \leqslant \infty$.

4. Every Lie group is parallelizable. (A *Lie group* is a manifold G together with a group operation $G \times G \to G$ which is C^ω, and such that inversion $G \to G$ is C^ω.)

2. Constructions with Vector Bundles

In this section we fix a differentiability class r, $0 \leqslant r \leqslant \omega$, and work consistently in the C^r category. For $r = 0$ this means we deal with topological spaces and continuous maps, while for $r > 0$ we deal only with C^r manifolds, C^r maps, and C^r vector bundles. Except for restrictions as indicated below, r is arbitrary. We write "bundle" for "vector bundle."

There is a general procedure, described in Lang's book [1], which for each functorial construction with vector spaces (direct sum, tensor product, etc.) defines a corresponding construction with vector bundles by applying the original construction to fibres. Rather than proceed at this level of abstraction, we describe explicitly the constructions we shall need.

A *subbundle* of a bundle $\xi = (p,E,B)$ is a bundle $\xi_0 = (p_0,E_0,B)$ over the same base space B, such that $E_0 \subset E$, $p_0 = p|E_0$, and there exists a vector bundle atlas Φ for ξ with the following property. There is a linear subspace of $\mathbb{R}^n$, which we may take to be $\mathbb{R}^k$, such that if $(\varphi,U) \in \Phi$ then φ maps $p^{-1}(U) \cap E$ into $\mathbb{R}^k$, and the pair

$$(\varphi|p^{-1}(U) \cap E, U)$$

belongs to the vector bundle structure of ξ_0.

The notion of subbundle is patterned after the definition of submanifold; and in fact if $A \subset M$ is a C^{r+1} submanifold then TA is a C^r subbundle of T_AM.

If ξ_0 is a subbundle of ξ then the inclusion map $E_0 \to E$ is a monomorphism $\xi_0 \to \xi$ over 1_B. Conversely, in analogy with Theorem 1.3.1, if *η is a bundle over B and $F: \eta \to \xi$ is a monomorphism over 1_B then $F(\eta)$, with the bundle structure induced by F, is a subbundle of ξ*. It suffices to prove a local result; hence we may suppose ξ and η are the trivial bundles $B \times \mathbb{R}^n$ and $B \times \mathbb{R}^k$, $k \leqslant n$. The monomorphism $F: B \times \mathbb{R}^k \to B \times \mathbb{R}^n$ has the form

$$F(x,y) = (x, G_x(y))$$

where $F: B \to L(\mathbb{R}^k, \mathbb{R}^n)$ is of class C^r, and each linear map $F_x: \mathbb{R}^k \to \mathbb{R}^n$ is injective. Fix $x \in B$; put $F_x(\mathbb{R}^k) = E \subset \mathbb{R}^n$. There is no loss of generality in assuming that $E = \mathbb{R}^k \subset \mathbb{R}^n$ and F_x is the standard inclusion $\mathbb{R}^k \to \mathbb{R}^n$. Let

$\pi:\mathbb{R}^n \to \mathbb{R}^k$ be orthogonal projection. There is an open neighborhood $U \subset B$ of x such that $\pi F_z:\mathbb{R}^k \to \mathbb{R}^k$ is an isomorphism for $z \in U$. Let $K \subset \mathbb{R}^n$ be the kernel of π, so that $\mathbb{R}^n = \mathbb{R}^k \times K$. Define

$$\varphi: U \times (\mathbb{R}^k \times K) \to U \times (\mathbb{R}^k \times K)$$

by

$$\varphi(z,(v,w)) = (z,\pi G_z(v),w).$$

Then φ is a C^r vector bundle chart for ξ and φ takes the image of F into $U \times \mathbb{R}^k$. This shows that the image of F is a subbundle.

Another way of getting subbundles is to take the *kernel* of an epimorphism $F:\xi \to \xi'$ which covers 1_B. That is, for each $x \in B$ let η_x be the kernel of $F_x:\xi_x \to \xi'_x$; then there is a unique subbundle η of ξ having fibres η_x. We leave this for the reader to prove.

It is useful to introduce the notion of an *exact sequence* of vector bundles morphisms: this means a finite or infinite sequence

$$\cdots \to \xi_{i-1} \xrightarrow{F_{i-1}} \xi_i \xrightarrow{F_i} \xi_{i+1} \to \cdots$$

of morphisms, all covering 1_B, such that for each $x \in B$ we have

$$\text{image}\,(F_{i-1})_x = \text{kernel}\,(F_i)_x$$

for all i. Of particular interest are the *short* exact sequences

$$0 \to \xi \xrightarrow{F} \eta \xrightarrow{G} \zeta \to 0$$

where 0 denotes a 0-dimensional bundle over B. Such a sequence means merely that F is a monomorphism, G is an epimorphism and image F = kernel G.

The existence of kernel subbundles for epimorphism can be stated in functorial language; given the exact sequence

$$\eta \xrightarrow{G} \zeta \to 0$$

there is an exact sequence

$$0 \to \xi \xrightarrow{F} \eta \xrightarrow{G} \zeta \to 0 \tag{1}$$

and (1) is unique in the sense that for any exact sequence

$$0 \to \xi' \xrightarrow{F'} \eta \xrightarrow{G} \zeta \to 0$$

there is a unique isomorphism $\xi \to \xi'$ such that the diagram

$$\begin{array}{ccccccccc} 0 & \to & \xi & \xrightarrow{F} & \eta & \xrightarrow{G} & \zeta & \to & 0 \\ \downarrow & & & & \downarrow = & & \downarrow = & & \\ 0 & \to & \xi' & \xrightarrow[F']{} & \eta & \xrightarrow[G]{} & \zeta & \to & 0 \end{array}$$

commutes.

In the exact sequence (1) we call ζ the *quotient bundle* of the monomorphism F. It is easy to see that every monomorphism has a quotient bundle and the latter is unique up to isomorphism. In particular, if $\xi \subset \eta$

is a subbundle, the fibres of the quotient bundle are taken to be the vector spaces η_x/ξ_x, and we denote the quotient bundle by η/ξ.

The short exact sequence (1) is said to *split* if there is a monomorphism $H:\zeta \to \eta$ such that $GH = 1_\zeta$. Working fibrewise we see that this is equivalent to the existence of an epimorphism $k:\eta \to \xi$ such that $KF = 1_\xi$.

The *Whitney sum* (or *direct sum*) of bundles ξ, ζ over B is the bundle $\xi \oplus \zeta$ whose fibre over x is $\xi_x \oplus \zeta_x$. If φ, ψ are charts for ξ, ζ respectively over U, a chart θ for $\xi \oplus \zeta$ over U is obtained by setting

$$\theta_x = \varphi_x \oplus \psi_x : \xi_x \oplus \zeta_x \to \mathbb{R}^m \oplus \mathbb{R}^n.$$

The natural exact sequences of vector spaces

$$0 \to \xi_x \xrightarrow{F_x} \xi_x \oplus \zeta_x \xrightarrow{G_x} \zeta_x \to 0$$

fit together to give a split exact sequence

$$0 \to \xi \xrightarrow{F} \xi \oplus \zeta \xrightarrow{G} \zeta \to 0.$$

Let $\xi = (p,E,M)$ be a C^{r+1} vector bundle. Each fibre ξ_x is a vector space, with origin x; hence we identify ξ_x with $T_x(\xi_x)$. Thus ξ is a subbundle of $T_M E$ in a natural way. (Note that the "natural" differentiability class of $T_M E$ is only C^r.) Since $M \subset E$ is a submanifold (via the zero section), TM is a C^r subbundle of $T_M E$. Evidently we have a short exact sequence

$$(2) \qquad 0 \to \xi \to T_M E \xrightarrow{Tp} TM \to 0$$

which is split by

$$(3) \qquad Ti : TM \to T_M E.$$

This proves:

2.1. Theorem. *Let $\xi = (p,E,M)$ be a C^{r+1} vector bundle, $0 \leqslant r \leqslant \omega$. The exact sequence (2) of C^r vector bundles is naturally split by (3). Thus there is a natural C^r isomorphism*

$$h_\xi : T_M E \approx \xi \oplus TM.$$

In particular $\xi \subset T_M E$ as a natural subbundle.

Here *natural* means with respect to C^{r+1} morphisms. If

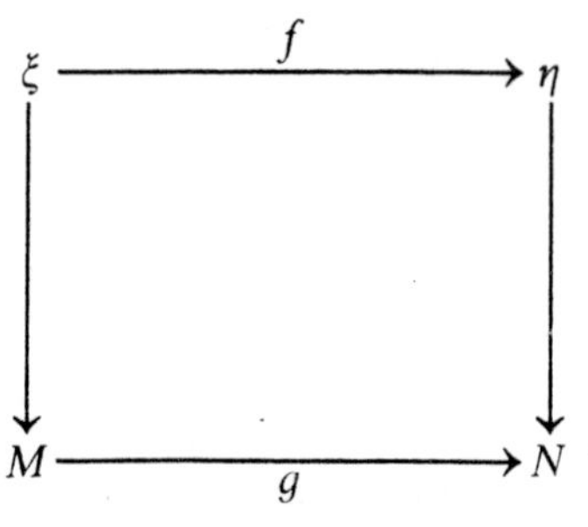

is such a morphism, then the diagram

$$\begin{array}{ccc} T_M E\xi & \xrightarrow{Tf} & T_N E\eta \\ \downarrow h_\xi & & \downarrow h_\eta \\ \xi \oplus TM & \xrightarrow[f \oplus Tg]{} & \eta \oplus TN \end{array}$$

commutes, as is easily checked.

The following simple result is one of the most useful facts about vector bundles.

2.2. Theorem. *Every short exact sequence of C^r vector bundles is split, $0 \leqslant r \leqslant \infty$, provided the base space is paracompact.*

Proof. It suffices to prove that a monomorphism $F:\xi \to \eta$ over 1_B has a left inverse. This is true locally because $F(\xi)$ is a subbundle of η; we showed above that there are charts for η covering B taking $(\eta|U,\xi|U)$ to $(U \times \mathbb{R}^n, U \times \mathbb{R}^k)$, and a local left inverse is obtained from a linear retraction $\mathbb{R}^n \to \mathbb{R}^k$. Since local left inverses can be glued together by a partition of unity, the theorem follows.

QED

Let $\xi = (p,E,B)$ be a vector bundle. An *inner product* or *orthogonal structure* (of class C^r) on ξ is a family $\alpha = \{\alpha_x\}_{x \in B}$ where each α_x is an inner product (symmetric, bilinear, positive definite 2-form) on the vector space E_x, such that the map $(x,y,z) \mapsto \alpha_x(y,z)$, defined on $\{(x,y,z) \in B \times E \times E: x = p(y) = p(z)\}$, is C^r. It is easy to construct such an α whenever B is paracompact and $r \leqslant \infty$, using partitions of unity. In fact, any K-germ of an orthogonal structure, where $K \subset B$ is a closed set, can be extended to an orthogonal structure. The pair (ξ,α) is called an *orthogonal vector bundle*. If M is a C^{r+1} manifold, a C^r orthogonal structure on TM is also called a *Riemannian metric* on M of class C^r.

Suppose (ξ,α) is an orthogonal bundle. If y, z are in the same fibre ξ_x we write $\langle y,z \rangle$ or $\langle y,z \rangle_x$ for $\alpha_x(y,z)$. If $\eta \subset \xi$ is a subbundle, the *orthogonal complement* $\eta^\perp \subset \xi$ is the subbundle defined fibrewise by

$$(\eta^\perp)_x = (\eta_x)^\perp = \{y \in \xi_x : \langle y,z \rangle = 0, \text{ all } z \in \xi_x\}.$$

The natural epimorphism $\xi \to \xi/\eta$ maps $\eta^\perp$ isomorphically onto ξ/η. This provides another method of splitting short exact sequences, one which works just as well for analytic bundles with analytic inner products.

Let $M \subset N$ be a C^{r+1} submanifold; suppose N has a C^r Riemannian metric. In this case $TM^\perp \subset T_M N$ is called the *geometric normal bundle* of

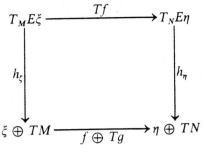

M in N. The *algebraic normal bundle* of M in N is the C^r quotient bundle $T_M N/TM$; it is canonically C^r isomorphic to TM .

Let $\xi = (p,E,M)$ be an orthogonal bundle. In each fibre ξ_x an orthonormal basis e_x can be derived from arbitrary basis b_x of ξ_x by the Gram-Schmidt orthogonalization method. This classical procedure, which is a deformation retraction of $GL(n)$ into $O(n)$, is so canonical that it leads to a C^r family of orthonormal bases $\{e_x\}_{x \in U}$, $U \subset M$, if one starts from an arbitrary C^r family $\{b_x\}_{x \in U}$. This shows that ξ has an *orthogonal atlas* $\Phi = \{\varphi_i, U_i\}$, that is, each map

$$\varphi_{ix}: \xi_x \to \mathbb{R}^n, \qquad x \in U_i$$

is an isometry. It follows that the transition functions

$$g_{ij}: U_i \cap U_j \to GL(n)$$

take values in $O(n)$. In other words every orthogonal bundle has an orthogonal atlas. Conversely, given an orthogonal atlas on ξ there is a unique orthogonal structure on ξ making each φ_{ix} isometric.

Two orthogonal vector bundles are *isomorphic* if there is a vector bundle isomorphism between them which preserves inner products.

The following lemma shows that the orthogonal structure on a vector bundle is essentially unique.

2.3. Lemma. *Let $\xi_i = (p_i,E_i,M)$ be a vector bundle, $i = 0, 1$ and f: $\xi_0 \to \xi_1$ an isomorphism. Suppose ξ_0 and ξ_1 have orthogonal structures. Then f is homotopic through vector bundle maps to an isomorphism of orthogonal bundles.*

Proof. Suppose first ξ_0 and ξ_1 are trivial as orthogonal bundles. We have

$$f: M \times \mathbb{R}^n \to M \times \mathbb{R}^n$$
$$f(x,y) = (x, g(x)y),$$
$$g: M \to GL(n).$$

Since $O(n)$ is a deformation retract of $GL(n)$, g is homotopic to $h: M \to O(n)$. Moreover the homotopy can be chosen rel $g^{-1}O(n)$. Writing such a homotopy as g_t, $0 \leqslant t \leqslant 1$, with $g_0 = g$, $g_1 = h$, we define

$$f_t: M \times \mathbb{R}^n \to M \times \mathbb{R}^n, \qquad 0 \leqslant t \leqslant 1$$
$$f_t(x,y) = (x, g_t(x)y).$$

This is a homotopy of vector bundle isomorphisms from f to an isomorphism of orthogonal bundles, and $f_t = f$ whenever f is already orthogonal.

The general case of Theorem 2.3 is proved by applying this special case successively over each element of a locally finite open cover $\{U_j\}$ of M such that ξ_0 and ξ_1 are trivial orthogonal bundles over U_j.

QED

A quite different construction is that of the induced bundle. Let $\xi = (p,E,M,\Phi)$ be a vector bundle and $f: M_0 \to M$ a map. The *induced bundle* (or *pullback*) $f^*\xi = (p_0,E_0,M_0,\Phi_0)$ is defined as follows. Put

$$E_0 = \{(x,y) \in M_0 \times E : f(x) = p(y)\},$$

and define

$$p_0 : E_0 \to M_0,$$
$$p_0(x,y) = x.$$

Take Φ_0 to be the maximal (C^r) atlas containing all charts of the form $(\psi, f^{-1}U)$ where $(\varphi,U) \in \Phi$ and if $z \in f^{-1}(U)$, $f(z) = x \in U$, then $\psi_z = \varphi_x$. The *natural vector bundle map* $\Psi : f^*\xi \to \xi$ over f is given by $(x,y) \mapsto y$.

Let $q:\eta \to M_0$ be a vector bundle and $F:\eta \to \xi$ a morphism over f. There is a unique map of total spaces $H:\eta \to f^*\xi$ making a commutative diagram

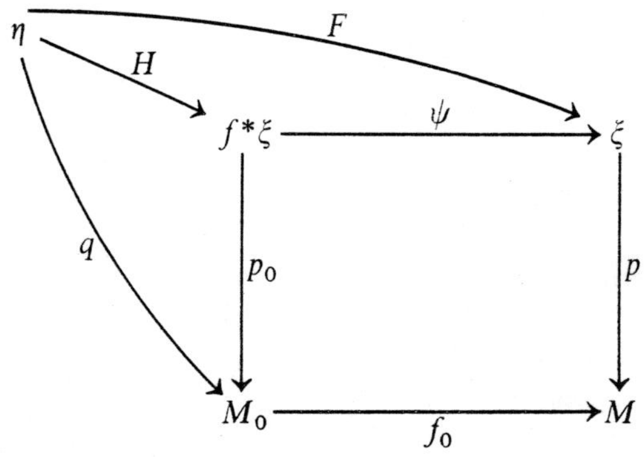

and H is an isomorphism of vector bundles. If F is an epimorphism, monomorphism, or bimorphism, so is H. This proves the useful fact that *if $F:\eta \to \xi$ is a vector bundle map over f then η is canonically isomorphic to the pullback $f^*\xi$.*

The main theorem about induced bundles is the following corollary of the covering homotopy Theorem 1.5.

2.4. Theorem. *Suppose B is a paracompact space. Let f, $g:B \to M$ be homotopic maps, and ξ a vector bundle over M. Then $f^*\xi$ is isomorphic to $g^*\xi$. In particular, if g is constant then $f^*\xi$ is trivial.*

Proof. Let $H:B \times I \to M$ be a homotopy from f to g. By 1.3, $H^*\xi$ is isomorphic to $(H_0^*\xi) \times I = (f^*\xi) \times I$ and also to $(H_1^*\xi) \times I = (g^*\xi) \times I$. Looking at these bundles over $B \times I$ we find that $f^*\xi$ is isomorphic to $g^*\xi$.

QED

2.5. Corollary. *Every vector bundle over a contractible paracompact space is trivial.*

The following corollary of 2.5 explains the name "covering homotopy theorem":

2.6. Theorem. *Let B_0 be paracompact. Let $F:\xi_0 \to \xi_1$ be a morphism of vector bundles covering $f:B_0 \to B_1$. Let $h:B_0 \times I \to B_1$ be a homotopy of f. Then there is a morphism $H:\xi_0 \times I \to \xi_1$ of F covering h. If F is a mono-, epi- or biomorphism, H has the same property.*

Proof. It suffices to find a morphism $\xi_0 \times I \to \xi_1 \times I$ which covers the map $g:B_0 \times I \to B_1 \times I$, $g(x,t) = (h(x,t),t)$ and which is given by F over $B_0 \times 0$. Using the relation between induced bundles and bundle maps, we see that we may replace $\xi_1 \times I$ with $g^*(\xi_1 \times I)$, g with the identity map of $B_0 \times I$, and $B_1 \times I$ with $B_0 \times I$. By 1.3 $g^*(\xi_1 \times I)$ can be replaced by a vector bundle $\eta \times I$ over $B_0 \times I$. Thus we are given an isomorphism map $F:\xi_0 \to \eta$ and we must extend it to an isomorphism $H:\xi_0 \times I \to \eta \times I$. We can take $H = F \times 1_I$. The last statement is obvious.

QED

Exercises

1. Let $\xi_i = (p_i,E_i,B_i)$ be a C^r vector bundle, $i = 0, 1$, and $f:B_0 \to B_i$ a C^r map, $0 \leqslant r \leqslant \omega$. There is a C^r vector bundle η over B_0 whose fibre over x is $L(\xi_{0x},\xi_{1x})$, such that C^r sections of η correspond naturally to C^r morphisms $\xi_0 \to \xi_1$ over f.

2. Let P^k denote real projective k-space, ε the trivial 1-dimensional vector bundle over P^k and ξ the normal bundle of $P^k \subset P^{k+1}$. Then $\varepsilon^1 \oplus TP^k \cong \xi \oplus \cdots \oplus \xi$. [Consider the inclusion $S^k \subset S^{k+1}$ and the antipodal map.]

3. (a) If n is odd TS^n has a nonvanishing section and therefore $TS^n = \varepsilon^1 \oplus \eta$ where ε^k is a trivial k-dimensional bundle. [If $n = 2m - 1$, $S^n \subset \mathbb{R}^{2m} = \mathbb{C}^m$. If $x \in S^n_x$ and $i = \sqrt{-1}$ then ix is tangent to S^m at x.]

(b) If $n = 4m - 1$, $TS^n \cong \varepsilon^3 \oplus \zeta$. [Use quaternions.]

(c) If $n = 8m - 1$, $TS^n \cong \varepsilon^7 \oplus \lambda$ [Use Cayley numbers.]

4. $TS^n \oplus \varepsilon^1$ is trivial. [Consider $T(S^n \times \mathbb{R}) \subset T(\mathbb{R}^{n+1})$. Compare Exercise 12 of Section 1.2.]

5. If TM has a nonvanishing section and $TM \oplus \varepsilon^1$ and $TN \oplus \varepsilon^1$ are trivial then $M \times N$ is parallelizable.

6. A product of two or more spheres is parallelizable if they all have positive dimensions and at least one has odd dimension. [Use Exercises 3(a), 4, and 5, and induction.]

***7.** Find explicit trivializations of the tangent bundles of $S^1 \times S^2$, $S^1 \times S^4$, $S^2 \times S^5$.

8. The *frame bundle* $F(M)$ of an n-manifold M is the manifold of dimension $n^2 + n$ whose elements are the pairs (x,λ) where $x \in M$ and $\lambda:\mathbb{R}^n \to M_x$ is a linear isomorphism. Define $\pi:F(M) \to M$ by $\pi(x,\lambda) = x$. The topology and differential structure on $F(M)$ are such that a coordinate system (φ,U) on M induces a diffeomorphism $\pi^{-1}(U) \approx U \times GL(n)$ by $(x,\lambda) \mapsto (x,D\varphi_x \circ \lambda)$.

(a) There is an exact sequence

(S) $$0 \to \text{kernel}\,(T\pi) \to TF(M) \to \pi^*(TM) \to 0.$$

(b) kernel $(T\pi)$ and $\pi^*(TM)$ are trivial vector bundles.

(c) Therefore $F(M)$ is parallelizable.

9. (a) Parts (a) and (b) of Exercise 8 are true even if M is not paracompact. In this case, however, the exact sequence (S) might not split. A splitting $j: \pi^*(TM) \to TF(M)$ of (S) is called a (perhaps nonlinear) *connection* on M.

(b) Let M be a connected manifold, not assumed to be paracompact or to have a countable base. If M has a connection then M is paracompact and has a countable base. [Hint: use part (a) and Exercise 8(c) to get a Riemannian metric on $F(M)$.]

3. The Classification of Vector Bundles

We now prove a basic result—ultimately based on transversality—which quickly leads to the classification Theorem 3.4.

3.1. Theorem. *Let ξ be a k-dimensional C^r vector bundle over a manifold M, $0 \leqslant r \leqslant \infty$. Let $U \subset M$ be a neighborhood of a closed set $A \subset M$. Suppose that*

$$F: \xi|U \to U \times \mathbb{R}^s$$

is a C^r monomorphism (of vector bundles) over 1_U. If $s \geqslant k + \dim M$ then there is a C^r monomorphism $\xi \to M \times \mathbb{R}^s$ over 1_M which agrees with F over some neighborhood of A in U.

Proof. Consider first the special case where ξ is trivial. Then for each $x \in U$, $F|\xi_x$ is a linear map $g(x): \mathbb{R}^k \to \mathbb{R}^s$ of rank k. We thus obtain a map $g: U \to V_{s,k}$, which is easily proved to be C^r. Since $\dim M \leqslant s - k$ we can apply Theorem 3.2.5 (on extending continuous maps into $V_{n,k}$) to find a C^r map $h: M \to V_{s,k}$ which extends the A-germ of g. If $r > 0$ we use the relative approximation Theorem 2.2.5 to make h C^r. Then we interpret h as a monomorphism $M \times \mathbb{R}^k \to M \times \mathbb{R}^s$ over 1_M.

The general case follows by using the globalization Theorem 2.2.11.

(For those who want more details: a structure functor $(\mathscr{F}, \mathfrak{Y})$ on M is defined as follows. Let $\mathscr{X} = \{X_i\}$ be a locally finite closed cover of M such that ξ is trivial over a neighborhood of each Y_i. Let $\mathfrak{Y}$ be the family of all unions of elements of $\mathscr{X}$. For $Y \in \mathfrak{Y}$ let $\mathscr{F}(Y)$ be the set of equivalence classes $[\varphi]_Y$ ("Y-germs") of maps $\varphi: \xi|W \to W \times \mathbb{R}^s$, as follows. $W \subset M$ can be any neighborhood of Y, and φ must be a monomorphism over 1_W which agrees with f over some neighborhood of $Y \cap A$. The equivalence relation is: $[\varphi]_Y = [\psi]_Y$ if φ and ψ agree over some neighborhood of Y. Restriction makes $(\mathscr{F}, \mathfrak{Y})$ into a structure functor which is clearly continuous and nontrivial; and we proved above that $(\mathscr{F}, \mathfrak{Y})$ is locally extendable. Therefore Theorem 2.2.11 yields Theorem 3.1.)

QED

Let $\gamma_{s,k} \to G_{s,k}$ be the following vector bundle over the Grassmannian $G_{s,k}$: the fibre of $\gamma_{s,k}$ over the k-plane $P \subset \mathbb{R}^s$ is the set of pairs (P,x) where $x \in P$. This makes $\gamma_{s,k}$ into an analytic k-dimensional vector bundle in a natural way. We call this the *Grassmann bundle* or sometimes the *universal bundle* over $G_{s,k}$.

From Theorem 3.1 we have:

3.2. Corollary. *Let η be a C^r k-plane bundle over $V \times I$ where V is an n-manifold. Suppose that $F_i:\eta|V \times i \to \mathbb{R}^s$ is a C^r monomorphism for $i = 0, 1$. If $s > k + n$ then F_0 and F_1 extend to a C^r monomorphism $F:\eta \to \mathbb{R}^s$.*

Proof. By the covering homotopy theorem, F_0 and F_1 extend to a monomorphism $\eta|U \to \mathbb{R}^s$, where $U \subset V \times I$ is a neighborhood of $V \times \{0,1\}$. Now apply 3.1 (with $M = V \times I$, $A = V \times \{0,1\}$, etc.)

QED

Another corollary of Theorem 3.1 is the existence of "inverse" bundles:

3.3. Theorem. *Let ξ be a C^r k-plane over an n-manifold M, $0 \leqslant r \leqslant \infty$. Then there is a C^r n-plane bundle η over M such that $\xi \oplus \eta \cong_r M \times \mathbb{R}^s$.*

Proof. Let $F:\xi \to M \times \mathbb{R}^{n+k}$ be a monomorphism over 1_M. Give the trivial bundle $M \times \mathbb{R}^{n+k}$ its standard orthogonal structure, and for η take subbundle $F(\xi)^\perp \subset M \times \mathbb{R}^{n+k}$.

QED

We now give another meaning to a C^r monomorphism $F:\xi \to M \times \mathbb{R}^s$ over 1_M. Define a vector bundle map

$$\begin{array}{ccc} \xi & \xrightarrow{\varphi} & \gamma_{s,k} \\ \downarrow & & \downarrow \\ M & \xrightarrow[g]{} & G_{s,k} \end{array} \tag{1}$$

as follows. To $y \in M$, g assigns the k-plane $g(y) = F(\xi_y) \in G_{s,k}$. If $z \in \xi_y$ defines $\varphi(x) = (F(\xi_y), f(z))$. It is easy to see that this correspondence $F \mapsto (\varphi, g)$ induces a natural bijection between C^r monomorphisms $\xi \to M \times \mathbb{R}^s$ over 1_M, and C^r bundle maps $\xi \to \gamma_{s,k}$.

The map $g:M \to G_{s,k}$ in diagram (1) has the property that $g^*\gamma_{s,k} \cong \xi$. Such a map is called a *classifying map* for ξ; we also say *g classifies* ξ. From Theorems 3.1 and 3.2 and a collar on ∂M (see Section 4.6), we obtain the following *classication theorem*:

3.4. Theorem. *If $s \geqslant k + n$ then every C^r k-plane bundle ξ over an n-manifold M has a classifying map $f_\xi:M \to G_{s,k}$. In fact any classifying map $\partial M \to G_{s,k}$ for $\xi|\partial M$ extends to a classifying map for ξ. When $s > k + n$ the homotopy class of f_ξ is unique, and if η is another k-plane bundle over M then $f_\xi \simeq f_\eta$ if and only if $\xi \cong \eta$.*

Proof. The only statement needing further proof is the "if" clause. Suppose $\psi:\eta \cong \xi$ and let $\varphi:\xi \to \gamma_{s,k}$ be a bimorphism covering f_ξ. From

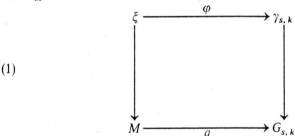

the diagram

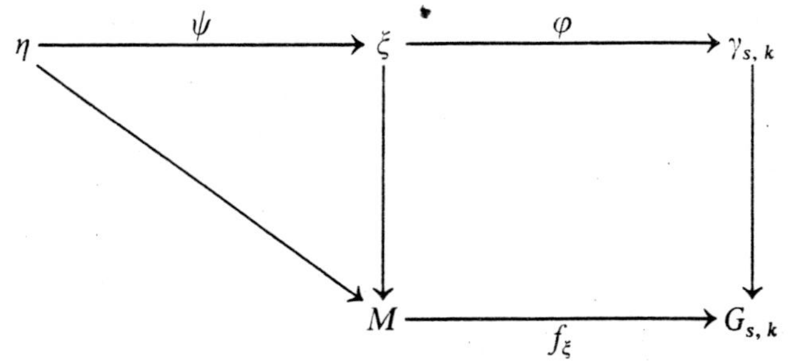

we see that f_ξ classifies *both* ξ and η. Since the classifying map f_η is unique up to homotopy when $s > k + n$ it follows that $f_\xi \simeq f_\eta$.

QED

Taken together with the covering homotopy theorem, this result is of fundamental importance because it converts the theory of vector bundles into a branch of homotopy theory. To put it another way, we can use whatever we know about maps to study vector bundles. For example, approximation theory yields:

3.5. Theorem. *Every C^r vector bundle ξ over a C^∞ manifold M has a compatible C^∞ bundle structure; and such a structure is unique up to C^∞ isomorphism.*

Proof. Let $g: M \to G_{s,k}$ be a C^r classifying map for ξ. Then g can be approximated by, and so is homotopic to, a C^∞ map h. Therefore

$$\xi \cong_r g^*\gamma_{s,k} \cong_r h^*\gamma_{s,k}.$$

But $h^*\gamma_{s,k}$ is a C^∞ bundle. Thus ξ has a C^∞ structure.

If η_0 and η_1 are C^∞ bundles that are C^r isomorphic, they have C^∞ classifying maps that are homotopic. These maps are then C^∞ homotopic. Pulling back $\gamma_{s,k}$ over $M \times I$ by such a homotopy gives a C^∞ vector bundle ξ such that

$$\zeta | M \times i \cong_\infty \eta_i, \qquad i = 0.$$

Therefore $\eta_0 \cong_\infty \eta_1$ by Theorem 1.4.

QED

The same result is true if C^∞ is replaced by C^ω; the proof uses the analytic approximation Theorem 2.5.1. See also Exercise 3 of Section 4.7 for a theorem of this type that can be proved without using Theorem 2.5.1.

From now on we need not specify the differentiability class of a vector bundle.

Although the theorems of this section have been stated for manifolds, they are also true (ignoring differentiability) for vector bundles over simplicial (or CW) complexes of finite dimension. The proofs are almost the same. The main difference is that Theorem 3.1 is proved by induction on dimension;

the inductive step is proved by extending a map $\partial\Delta^m \to V_{s,k}$ to Δ^m if $m \leqslant s - k$, where Δ^m is an m-simplex. Similarly for CW complexes.

The classification of vector bundles over more general spaces can be stated as follows. Let $K^k(X)$ denote the set of isomorphism classes of k-plane bundles over the space X. Let $[X,G_{s,k}]$ be the set of homotopy classes of maps from X to $G_{s,k}$. A natural map

$$\Theta_X : K^k(X) \to [X,G_{s,k}]$$

is induced by the correspondence $f \mapsto f^*\gamma_{s,k}$; that Θ_X is well defined follows from the covering homotopy theorem, if we assume X paracompact. Then one can prove: *if X has the homotopy type of a simplicial or CW complex of dimension less than $s - k$, the map Θ_X is bijective.*

Exercises

[X denotes either a manifold, or a finite dimensional simplicial or CW complex; ε^k denotes the trivial k-plane bundle.]

1. (a) Let $i: G_{s,k} \to G_{s+1,k+1}$ be the natural inclusion. Then

$$i^*\gamma_{s+1,k+1} \cong \gamma_{s,k} \oplus \varepsilon^1.$$

(b) If $\dim X < s - k$, then under the classification of vector bundles the map

$$i_\# : [X,G_{s,k}] \to [X,G_{s+1,k+1}]$$

corresponds to the map

$$\sigma : K^kX \to K^{k+1}X, \qquad [\xi] \mapsto [\xi \oplus \varepsilon].$$

2. Suppose $\dim X < \min\{s - k, r - j\}$. The natural embedding

$$G_{s,k} \times G_{r,j} \to G_{s+r,k+j}$$

induces the map

$$K^kX \times K^jX \to K^{k+j}X$$

which corresponds to Whitney sum.

3. The map $\sigma: K^kX \to K^{k+1}X$ (see Exercise 1) is surjective if $\dim X \leqslant k$ and injective if $\dim X < k$. [Use Exercise 7a and Exercise 9, Section 3.2.]

4. Let ξ, η, ζ be bundles over X such that $\xi \oplus \eta \cong \xi \oplus \zeta$. If $\dim \xi > \dim X$ then $\eta \cong \zeta$. [Suppose $\eta \oplus \alpha$ is trivial; use Exercise 3.]

5. Let $G_{s,k} \to G_{s+1,k}$ be the natural inclusion and put $G_{\infty,k} = \bigcap_{s=k}^{\infty} G_{s,k}$. Then K^kX is naturally isomorphic to $[X,G_{\infty,k}]$. (More usually $G_{\infty,k}$ is denoted by or $BO(k)$. It is called the classifying space for the functor K^k, and also for the group $O(k)$.)

6. Two vector bundles ξ, η over X are *stably isomorphic* if $\xi \oplus \varepsilon^j \cong \eta \oplus \varepsilon^k$ for some j, k. Let KX denote the set of stable isomorphism classes of bundles over X. The operation of Whitney sum induces a natural abelian group structure on KX.

***7.** There are maps $G_{\infty,k} \to G_{\infty,k+1}$ whose direct limit $G_{\infty,\infty}$ is a classifying space for the functor KX of Exercise 6. That is, there is a natural isomorphism (of sets) $KX \approx [X,G_{\infty,\infty}]$. Moreover, there is a map $G_{\infty,\infty} \times G_{\infty,\infty} \to G_{\infty,\infty}$ such that the resulting binary operation on $[X,G_{\infty,\infty}]$ corresponds to the Whitney sum operation in KX. (More usually $G_{\infty,\infty}$ is denoted by BO.)

8. A k-plane bundle over S^n has a vector bundle atlas containing only two charts, each of whose domains contain a hemisphere. The transition function for such a pair of charts restricts to a map of the equator S^{n-1} into $GL(k)$. In this way an isomorphism $K^k(S^n) \approx \pi_{k-1}(GL(k)) \approx \pi_{k-1}(O(k))$ is established, with no restriction on k, n.

9. Let $\xi \to S^n$ be a k-plane bundle corresponding to $\alpha \in \pi_{n-1}(GL(k))$ (see Exercise 8). If $\eta \to S^n$ corresponds to the inverse of ξ (in the abelian group $\pi_{n-1}(GL(k))$) then $\xi \oplus \eta$ is trivial. One can interpret η as the "reflection" of ξ in the equator.

***10.** Every vector bundle over S^3 is trivial. [Hint: it suffices to consider 3-plane bundles.]

4. Oriented Vector Bundles

Let V be a (real) finite dimensional vector space of dimension $n > 0$. Two bases $(e_1, \ldots, e_n)$, $(f_1, \ldots, f_n)$ of V are *equivalent* if the automorphism $A: V \to V$ such that $Ae_i = f_i$ has positive determinant. An *orientation* of V is an equivalence class $[e_1, \ldots, e_n]$ of bases. If $\dim V > 0$ there are just two orientations. If one of them is denoted by ω, then $-\omega$ denotes the other one.

If $L: V \to W$ is an isomorphism of vector spaces and $\omega = [e_1, \ldots, e_n]$ is an orientation of V then $L(\omega) = [Le_1, \ldots, Le_n]$ is the *induced* orientation of W.

If $\dim V = 0$ an *orientation* of V simply means one of the numbers ± 1. Many special but trivial arguments for this case will be omitted.

An *oriented vector space* is a pair (V, ω) where ω is an orientation of V. Given (V, ω) and (V', ω') an isomorphism $L: V \to W$ is called *orientation preserving* if $L(\omega) = \omega'$; otherwise L is *orientation reversing*.

The *standard orientation* ω^n of $\mathbb{R}^n$, $n > 0$, is $[e_1, \ldots, e_n]$ where e_i is the i'th unit vector. The standard orientation of $\mathbb{R}^0$ is $+1$.

Let $0 \to E' \xrightarrow{\varphi} E \xrightarrow{\psi} E'' \to 0$ be an exact sequence of vector spaces. Given orientations $\omega' = [e_1, \ldots, e_m]$ of E' and $\omega'' = [f_1, \ldots, f_n]$ of E'', an orientation ω of E is defined by

$$\omega = [\varphi e_1, \ldots, \varphi e_m, g_1, \ldots, g_n] \qquad \text{where} \qquad \psi g_i = f_i,$$

independent of the choice of the g_i. For if also $\psi h_i = f_i$, the automorphism $A: E \to E$ such that $Ae_j = e_j$ and $Ag_i = h_i$ fits into the commutative diagram

$$\begin{array}{ccccccccc}
0 & \longrightarrow & E' & \xrightarrow{\varphi} & E & \xrightarrow{\psi} & E'' & \longrightarrow & 0 \\
 & & \Big\downarrow{=} & & \Big\downarrow{A} & & \Big\downarrow{=} & & \\
0 & \longrightarrow & E' & \xrightarrow[\varphi]{} & E & \xrightarrow[\psi]{} & E'' & \longrightarrow & 0
\end{array}$$

which implies that $\det A = 1$. Hence

$$[\varphi e_1, \ldots, \varphi e_m, g_1, \ldots, g_n] = [\varphi e_1, \ldots, \varphi e_m, h_1, \ldots, h_m].$$

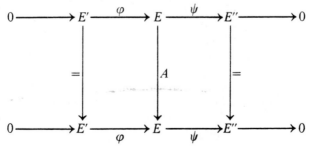

It is easy to see that any two of ω', ω, ω'' determine the third uniquely. We write $\omega = \omega' \oplus \omega''$, $\omega' = \omega/\omega''$, $\omega'' = \omega/\omega'$.

Now let $\xi = (p,E,B)$ be a vector bundle. An *orientation* for ξ is a family $\omega = \{\omega_x\}_{x \in B}$ where ω_x is an orientation of the fibre E_x such that ξ has an atlas Φ with the following property: if $\varphi:\xi|U \to \mathbb{R}^n$ is in Φ then

$$\varphi_x:(E_x,\omega_x) \to (\mathbb{R}^n,\omega_n)$$

is orientation preserving. We call ω a *coherent* family of orientations of the fibres. The atlas Φ is an *oriented atlas belonging* to ξ.

If ξ has an orientation ω, then ξ is called *orientable* and the pair (ξ,ω) is an *oriented vector bundle*. It is easy to see that if ξ is a C^r vector bundle for any $r \geqslant 1$, and ω is an orientation, then ξ has a C^r atlas which belongs to ω. An oriented C^r vector bundle can be defined as a C^r vector bundle together with a maximal C^r oriented atlas.

Let $F:\eta \to \xi$ be a bimorphism. If ξ has an orientation ω, there is a unique orientation θ of η such that F maps fibres to fibres preserving orientation. It follows that the pullback of $f^*\xi$ of an oriented bundle (ξ,ω) has a natural orientation $f^*\omega$.

Let $\xi = (p,E,B)$ be any vector bundle. Let $\lambda:I \to B$ be a path and ω an orientation of $\xi|\lambda(0)$. We propagate ω along λ as follows. Since the induced bundle $\lambda^*\xi \to I$ is trivial and I is connected, there is a unique orientation θ of $\lambda^*\xi$ such that over $0 \in I$, θ coincides with $\lambda^*\omega$. Denote by $\lambda_\#\omega$ the orientation of $\xi|\lambda(1)$ such that over $1 \in I$, $\lambda^*(\lambda_\#\omega)$ coincides with θ.

Let $\mu:I \to B$ be another path with $\lambda(0) = \mu(0)$ and $\lambda(1) = \mu(1)$. *If* $\lambda \simeq \mu$ *rel* $\{0,1\}$, *then* $\mu_\#\omega = \lambda_\#\omega$. To see this, let $f:D^2 \to B$ be such that $f = \lambda$ on the top semicircle $I_+ \subset \partial D^2$, and $f = \mu$ on the bottom semicircle $I_- \subset \partial D^2$. Since D^2 is contractible, $f^*\xi$ is trivial and therefore orientable. Since D^2 is connected, $f^*\xi$ has a unique orientation θ containing $f^*\omega = \lambda^*\omega = \mu^*\omega$. Over $1 \in I$, therefore, the orientations θ, $f^*\lambda_\#\omega$ and $f^*\mu_\#\omega$ all coincide. This implies that $\lambda_\#\omega = \mu_\#\omega$.

It follows that *every vector bundle over a simply connected manifold M is orientable.* To see this, pick a point $x_0 \in M$ and for each $y \in M$ let $\lambda_y:I \to M$ be a path joining x_0 to y (we may assume M path connected). Let ω be an arbitrary orientation of $\xi|x_0$ and define $\omega_y = \lambda_{y\#}\omega$. Since M is simply connected ω_y is independent of the choice of λ_y. If $\varphi:U \approx \mathbb{R}^n$(or $\mathbb{R}^n_+$) is a chart on M and $y \in U$, we can choose λ_z, $z \in U$ to be of the form λ_y followed by a straight line path (with respect to φ) from y to z. This choice of the λ_z shows that the resulting family of orientations $\{\omega_z\}_{z \in U}$ is an orientation of $\xi|U$. It follows that $\omega = \{\omega_y\}_{y \in M}$ is an orientation of ξ.

More generally, the vector bundle $\xi \to M$ is orientable if and only if every loop $\lambda:I \to M$, $\lambda(0) = \lambda(1)$, *preserves orientation* of $\xi|\lambda(0)$; that is, $\lambda_\#\omega = \omega$ if ω is an orientation of $\xi|\lambda(0)$. If this condition is satisfied and M is connected, then a given orientation of a single fibre ξ_x extends to a unique orientation of ξ by propagation along paths. Since each fibre has exactly two orientations, we see that an *orientable vector bundle over a connected manifold has just two orientations.* If one of these is called ω, the other is called $-\omega$.

In general every vector bundle $\xi = (p,E,B)$ has an oriented double covering $\tilde{\xi} = (\tilde{p},\tilde{E},\tilde{B})$. Let

$$\tilde{B} = \{(x,\omega): x \in B,\ \omega \text{ is an orientation of } \xi_x\}.$$

The topology of $\tilde{B}$ is generated by the subsets $\{x,\theta_x\}_{x \in U}$ where $U \subset B$ is open, $\xi|U$ is orientable, and θ is an orientation of $\xi|U$. There is a natural map $p: \tilde{B} \to B$, $p(x,\omega) = x$. Define $\tilde{\xi} = p^*\xi$. The *natural orientation of* $\tilde{\xi}$ is defined as follows: given $(x,\omega) \in \tilde{B}$, assign to $\xi|(x,\omega)$ the orientation $p^*\omega$. A section $B \to \tilde{B}$ is the same as an orientation of ξ. Therefore ξ is orientable if B is simply connected.

Let $0 \to \xi' \to \xi \to \xi'' \to 0$ be a short exact sequence of vector bundles. Given orientations ω', ω'' for ξ', ξ'' respectively, a family $\omega = \{\omega_x\}_{x \in B}$ of orientations of fibres of ξ is obtained by setting $\omega_x = \omega'_x \oplus \omega''_x$. Local trivializations make it clear that ω is coherent; thus ω is an orientation of ξ. Any two of ω, ω', ω'' determine the third. We put $\omega = \omega' \oplus \omega''$, etc. In particular we have

4.1. Lemma. *Two of ξ, ξ', ξ'' are orientable if and only if the third is.*

Let M be a manifold. M is called *orientable* if TM is an orientable vector bundle. An *orientation* of M means an orientation of TM; an *oriented manifold* is a pair (M,ω) where ω is an orientation of M. We define $-\omega$ to be the orientation of M such that $(-\omega)_x = -\omega_x$ (these are orientations of M_x) for all $x \in M$. If M is connected and orientable then it has exactly two orientations, ω and $-\omega$. Every simply connected manifold is orientable.

An alternative definition of "orientable manifold" is: M is orientable if it has an atlas whose coordinate changes have positive Jacobian determinants at all points. A maximal atlas of this kind is an *oriented differential structure.* By considering natural vector bundle charts it is easy to see that the two definitions are equivalent.

Let (M,ω) and (N,θ) be oriented manifolds. A diffeomorphism $f: M \approx N$ is called *orientation preserving* if $Tf: (TM,\omega) \to (TN,\theta)$ preserves orientation; in this case we write $f(\omega) = \theta$. On the other hand f is called *orientation reversing* if Tf reverses orientation. Notice that when M is connected, f must have one of these properties; to determine which one, it suffices to see whether a single $T_x f$ preserves orientation.

Now let M be a connected orientable manifold and let $g: M \approx M$ be a diffeomorphism. Let ω, $-\omega$ be the two orientations of M. Then either $g(\omega) = \omega$ and $g(-\omega) = -\omega$, or else $g(\omega) = -\omega$ and $g(-\omega) = \omega$. In other words g either preserves both orientations or reverses both orientations. We call g orientation preserving or orientation reversing, accordingly, independent of any choice of orientation for M. If Φ is an oriented differential structure for M then f preserves orientation if $f^*\Phi = \Phi$, that is, if the derivative of f at any point, expressed by charts in Φ, has positive Jacobian.

As an example consider a diffeomorphism of S^n obtained from an orthogonal linear operator $L \in O(n+1)$. Since L also maps D^{n+1} onto itself, $L|S^n$ is orientation-preserving exactly when $L|D^{n+1}$ is orientation-preserving, or equivalently, when L preserves orientation of $\mathbb{R}^{n+1}$. Thus *$L|S^n$ preserves orientation if and only if det $L > 0$.* In particular, *reflection in a hyperplane always reverses orientation; the antipodal map of S^n preserves orientation if n is odd and reverses orientation if n is even.*

Let M be a manifold and $\tilde{\xi} = (\tilde{p}, \tilde{E}, \tilde{M})$ the oriented double covering of the vector bundle TM. It is easy to see that $\tilde{\xi}$ is naturally isomorphic to $T\tilde{M}$. Therefore $\tilde{M}$ is an orientable manifold. This shows that *every manifold M has an oriented double covering $\tilde{M}$*. It is easy to see that the natural map $p: \tilde{M} \to M$ is a submersion. If M is orientable, then each orientation ω of M defines a section $s_\omega: M \to \tilde{M}$ by $s_\omega(x) = (x, \omega_x)$. Conversely, every section defines an orientation of M.

Next consider the algebraic normal bundle ν of ∂M in M:

4.2. Theorem. *ν is trivial, and hence orientable.*

Proof. Let $n = \dim M$. Put

$$\mathbb{R}^n_+ = \{x \in \mathbb{R}^n : x_1 \geqslant 0\}.$$

Let $\pi: \mathbb{R}^n \to \mathbb{R}$ be the projection $\pi(x) = x_1$.

Let $\{\varphi_i: U_i \to \mathbb{R}^n_+\}$ be family of charts of M that cover ∂M.

Define morphisms

$$F_i: TM|\partial U_i \to \mathbb{R},$$
$$F_{ix} = D(\pi\varphi_i)_x.$$

Since F_i maps $T(\partial U_i)$ to 0, it induces a morphism

$$G_i: \nu|\partial U_i \to \mathbb{R}$$

which is clearly a bimorphism. Note especially that if $x \in \partial U_i \cap \partial U_j$, the linear map

$$G_{jx}G_{ix}^{-1}: \mathbb{R} \to \mathbb{R}$$

is positive. This is equivalent to the fact that each φ_i maps U_i onto the same side of $\partial\mathbb{R}^n_+$ in $\mathbb{R}^n$. This already proves ν orientable. A trivialization of ν is obtained by gluing together the G_i with a partition of unity.

QED

Implicit in the last part of the above proof is this result:

4.3. Theorem. *An orientable 1-dimensional vector bundle over a paracompact space is trivial.*

This is true in all C^r categories; as usual the analytic case requires a separate proof. But Theorem 4.3 is false without paracompactness: the tangent bundle of the long line is orientable, but if it were trivial the long line would have a Riemannian metric and thus would be metrizable!

We now give some classical geometric applications of orientations.

4.4. Lemma. *Let N be a connected manifold and let $M \subset N$ be a connected closed submanifold of codimension 1, $\partial M = \partial N = \varnothing$. If M separates N then the normal bundle ν of M in N is trivial, and $N - M$ has exactly two components; and the (topological) boundary of each component is M.*

Proof. Let $A \subset W$ be a component of $N - M$. Then M is the boundary of the subset A. For, since M is closed, Bd A is a nonempty closed subset of M. Looking at submanifold charts for (N,M) one sees that Bd A is also open in M. Since M is connected, Bd $A = M$. Such charts also show that $\bar{A}$ is a submanifold of N with $\partial \bar{A} = M$. Clearly ν is also the normal bundle of M in $\bar{A}$; therefore Theorem 4.2 implies that ν is trivial. Let B be another component of $N - M$; then $\bar{B}$ is also a submanifold of N with boundary M. Thus $\bar{A} \cup \bar{B}$ is a closed subset of N. Invariance of domain (or the inverse function theorem) show that $\bar{A} \cup \bar{B}$ is also open. Therefore $\bar{A} \cup \bar{B} = N$.
From Theorems 4.1 and 4.4 we obtain:

QED

4.5. Theorem. *Let N be a simply connected manifold and $M \subset N$ a closed connected submanifold of codimension 1, $\partial M = \partial N = \varnothing$. If M separates N then M is orientable if N is orientable.*

Next, a basic topological result:

4.6. Theorem. *Let N be a simply connected manifold and $M \subset N$ a connected compact submanifold of codimension 1, $\partial M = \partial N = \varnothing$. Then M separates N.*

Proof. We may suppose N connected. Let $x_0, x_1 \in N - M$. Let $f: I \to N$ be a C^∞ path from $x_0 = f(0)$ to $x_1 = f(1)$; assume f is transverse to M. Then $f^{-1}(M)$ is a finite subset of I. Let $L(x_0,x_1,f) \in \mathbb{Z}_2$ be the reduction mod 2 of the cardinality of $f^{-1}(M)$. We assert that *$L(x_0,x_1,f)$ is independent of f*. For let $g: I \to M$ be another such path. Since N is simply connected, the paths f, g are homotopic rel end points. Thus there is a map $H: I \times I \to N$ such that $H(t,0) = f(t)$, $H(t,1) = g(t)$, $H(0,t) = x_0$, and $H(1,t) = x_1$. By approximation we may assume that H is C^∞ and transverse to M. Then $H^{-1}(M)$ is a compact 1-dimensional submanifold of $I \times I \subset \mathbb{R}^2$ with boundary $f^{-1}(0) \times 0 \cup g^{-1}(0) \times 1$. Since $H^{-1}(M)$ has an even number of boundary points, the assertion follows.

It is clear that there exist x_0, x_1, f as above with $L(x_0,x_1,f) = 1$; for example, take x_0 and x_1 on opposite sides of M, in a small arc transverse to M. Then x_0 and x_1 must be in different components of $N - M$, since otherwise there would exist a path g joining them in $N - M$. Such a path can be made C^∞ and transverse to N; then $L(x_0,x_1,f) = 0$, contradicting the assertion above.

QED

As a corollary of Theorem 4.6 we obtain the following "nonembedding theorem":

4.7. Theorem. *A compact nonorientable n-manifold without boundary cannot be embedded in a simply connected $(n + 1)$-manifold. In particular projective 2n-space P^{2n} does not embed in $\mathbb{R}^{2n+1}$ for $n \geq 1$.*

Proof. A simply connected manifold is orientable; together with Theorems 4.6 and 4.5, this proves the first statement. To prove the second we show that P^{2n} is nonorientable. Consider P^{2n} as the identification space of S^{2n} by the antipodal map A; let $p: S^{2n} \to P^{2n}$ be the projection. We know that A is orientation reversing. Therefore P^{2n} cannot be orientable; for if ω is an orientation of P^{2n}, there is a unique orientation θ of S^{2n} such that $T_x p(\theta_x) = \omega_{p(x)}$ for all $x \in S^{2n}$. But such a θ would be invariant under the antipodal map, which is impossible.

QED

Theorem 4.7 is false if "simply connected" is replaced by "orientable": for P^{2n} embeds in the orientable manifold P^{2n+1}.

It is also true that P^{2n+1} does not embed in R^{2n+2}, but more subtle methods are required.

Exercises

1. If ξ is any vector bundle, $\xi \oplus \xi$ is orientable. This implies that TM is orientable as a manifold.

2. There are precisely two isomorphism classes of n-plane bundles over S^1 for each $n \geq 1$. Two such bundles are isomorphic if and only if both are orientable or both are nonorientable.

3. $M \times N$ is orientable if and only if M and N are both orientable.

4. Every Lie group is an orientable manifold.

(In Exercises 5 through 9, $M \subset N$ is a closed, codimension 1 submanifold.)

5. If $\partial M \neq \varnothing$ and $\partial N = \varnothing$ and M, N are connected, then $N - M$ is connected.

6. Suppose $N = \mathbb{R}^{n+1}$, M is compact and $\partial M = \varnothing$. Then M bounds a unique compact submanifold of $\mathbb{R}^{n+1}$.

7. Suppose M is a neat submanifold. Then the normal bundle of M in N is trivial if and only if M has arbitrarily small neighborhoods in N that are separated by M.

8. Suppose M is compact and $\partial M = \varnothing$. If M is contractible to a point in N then M separates N.

9. If $M = \partial W$ where $W \subset N$ is a compact submanifold, and $W \neq N$, then M separates N.

10. Isomorphism classes of oriented k-plane bundles over an n-manifold M correspond naturally to homotopy classes of maps from M to the Grassman manifold $\tilde{G}_{s,k}$ of oriented k-planes in $\mathbb{R}^s$, provided $s > k + m$.

***11.** Let $\xi \to B$ be a nonorientable vector bundle over a connected manifold M. The set of homotopy classes of orientation-preserving loops at $x_0 \in M$ form a subgroup of index 2 in $\pi_1(M,x_0)$.

12. Let M be a connected orientable manifold. The subgroup $\mathrm{Diff}^r_+(M)$ of orientation preserving diffeomorphisms is normal and has index 1 or 2 in $\mathrm{Diff}^r(M)$, $1 \leqslant r \leqslant \infty$. Moreover $\mathrm{Diff}^r_+(M)$ is open and closed in both the weak and strong topologies.

5. Tubular Neighborhoods

Let $M \subset V$ be a submanifold. A *tubular neighborhood of* M (or *for* (V,M)) is a pair (f,ξ) where $\xi = (p,E,M)$ is a vector bundle over M and $f:E \to V$ is an embedding such that:

1. $f|M = 1_M$ where M is identified with the zero section of E;
2. $f(E)$ is an open neighborhood of M in V.

More loosely, we often refer to the open set $W = f(E)$ as a tubular neighborhood of M. It is then to be understood that associated to W is a particular retraction $q:W \to M$ making (q,W,M) a vector bundle whose zero section is the inclusion $M \to W$.

It is easy to see that only neat submanifolds can have tubular neighborhoods.

A slightly more general concept is that of a *partial tubular neighborhood* of M. This means a triple (f,ξ,U) where $\xi = (p,E,M)$ is a vector bundle over M, $U \subset E$ is a neighborhood of the zero section and $f:U \to V$ is an embedding such that $f|M = 1_M$ and $f(U)$ is open in V.

A partial tubular neighborhood (f,ξ,U) contains a tubular neighborhood, in the following sense: there is a tubular neighborhood (g,ξ) of M in V such that $g = f$ in a neighborhood of M.

To construct g, fix an orthogonal structure on ξ. Choose a map $\rho:M \to \mathbb{R}_+$ such that if $y \in E_x$ and $|y| \leqslant \rho(x)$ then $y \in f(U)$. Let $\lambda:[0,\infty) \to [0,1]$ be a diffeomorphism equal to the identity near 0. Define an embedding

$$h:E \to E,$$
$$h(y) = \rho(p|y|)\lambda(|y|)y.$$

Then $H(E) \subset U$ and $h =$ identity near M. Now put $g = fh$.

Eventually (Theorem 6.3) we shall prove that every neat submanifold has a tubular neighborhood. The first step is to prove:

5.1. Theorem. *Let* $M \subset \mathbb{R}^n$ *be a submanifold without boundary. Then* M *has a tubular neighborhood in* $\mathbb{R}^n$.

Proof. It suffices to find a partial tubular neighborhood.

Put $k = n - \dim M$ and let $\gamma_{n,k} \to G_{n,k}$ be the Grassmann bundle (see Section 4.3). Let $v:M \to G_{n,k}$ be a (C^∞) *field of transverse k-planes*; this means that for each $x \in M$, the tangent n-plane $M_x \subset \mathbb{R}^n$ is transverse to the k-plane $v(x)$. For example, one could take $v(x) = M_x^\perp$.

Put

$$\xi = (p,E,M) = v^*\gamma_{n,k};$$

thus ξ is a vector bundle, and

$$E = \{(x,y) \in M \times \mathbb{R}^n : y \in v(x)\}.$$

Define a map

$$f : E \to \mathbb{R}^n$$
$$f(x,y) = x + y \qquad (y \in v(x)).$$

The tangent space to E at a point $(x,0)$ of the zero section has a natural splitting $M_x \oplus v(x)$. It is clear that $T_{(x,0)} f$ is the identity on M_x and on $v(x)$. Therefore Tf has rank n at all points of the zero section and it follows that f is an immersion of some neighborhood of the zero section. Since $f|M = 1_M$ it follows that from Exercise 7, Section 2.1, $f|U$ is an embedding of some open neighborhood $U \subset E$ of M. Thus (f,ξ,U) is a partial tubular neighborhood of M.

QED

In the above construction, if we choose $v(x) = M_x^\perp$ the resulting tubular neighborhood is called a *normal tubular neighborhood* of M in $\mathbb{R}^n$. It is not hard to prove that in this case U can be chosen small enough so that $f(U \cap v_x)$ is the set of points in $f(U)$ whose nearest point of M is x. See Figure 4–1.

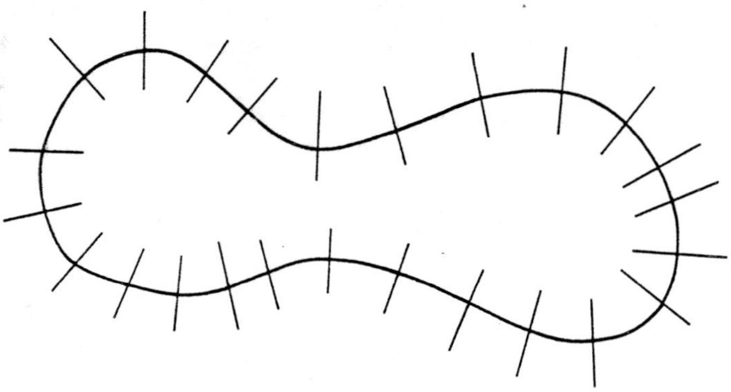

Figure 4–1. A normal tubular neighborhood.

5.2. Theorem. *Let $M \subset V$ be a submanifold, $\partial M = \partial V = \varnothing$. Then M has a tubular neighborhood in V.*

Proof. We may assume $V \subset \mathbb{R}^n$. Let $W \subset \mathbb{R}^n$ be a neighborhood of V and $r : W \to V$ a C^∞ retraction. (Such a W and r exist because V has a tubular neighborhood in $\mathbb{R}^n$.) Give V the Riemannian metric induced from $\mathbb{R}^n$ and let $v = (p,E,M)$ be the normal bundle of M in V. Thus

$$v \subset T_M V \subset T_M \mathbb{R}^n = M \times \mathbb{R}^n;$$

each fibre v_x is contained in $x \times \mathbb{R}^n$.

For each $x \in M$ let

$$U_x = \{(x,y) \in \nu_x : x + y \in W\}.$$

Put $U = \bigcup_{x \in M} U_x$. Then U is open in E, being the inverse image of W under the map

$$E \to \mathbb{R}^n,$$
$$(x,y) \mapsto x + y \qquad (y \in \nu_x).$$

It is easy to verify that the map

$$f : U \to V,$$
$$f(x,y) = r(x + y)$$

provides a partial tubular neighborhood for (V,M).

QED

It is useful to be able to slide one tubular neighborhood of a submanifold onto another one, mapping fibres linearly onto fibres. Such a sliding is a special case of an isotopy. Isotopies will be considered in more generality in a later chapter; at present the following remarks suffice.

If P, Q are manifolds, an *isotopy* of P in Q is a homotopy

$$F : P \times I \to Q,$$
$$F(x,t) = F_t(x)$$

such that the related map

$$\hat{F} : P \times I \to Q \times I,$$
$$(x,t) \mapsto (F_t(x),t)$$

is an embedding. We call $\hat{F}$ the *track* of F. We also say F is an isotopy *from* F_0 *to* F_1. If $A \subset P$ is such that $F_t(x) = F_0(x)$ for all $(x,t) \in A \times I$ then F is a *rel A isotopy*.

The relation "f is isotopic to g" is transitive. For let F, G be isotopies of P in Q such that $F_1 = G_0$. We can *almost* define an isotopy H from F_0 to G_1 by setting

$$H_t = \begin{cases} F_{2t} & 0 \leqslant t \leqslant \frac{1}{2} \\ G_{2t-1} & \frac{1}{2} \leqslant t \leqslant 2; \end{cases}$$

but H is not necessarily smooth at points of $P \times \frac{1}{2}$. The solution is to write instead:

$$H_t = \begin{cases} F_{2\tau(t)}, & 0 \leqslant t \leqslant \frac{1}{2}, \\ G_{2\tau(t)-1}, & \frac{1}{2} \leqslant t \leqslant 2 \end{cases}$$

where $\tau : I \to I$ is a C^∞ map which collapses a neighborhood of i to i for $i = 0, 1$. This H is indeed an isotopy from F_0 to G_1.

The same argument shows that rel A isotopy is an equivalence relation.

Now let $(f_i, \xi_i = (p_i, E_i, M))$ be a tubular neighborhood of $M \subset V$ for $i = 0, 1$. An *isotopy of tubular neighborhoods* from (f_0, ξ_0) to (f_1, ξ_1) is a rel

M isotopy F from E_0 to V such that:

$$F_0 = f_0,$$
$$F_0(E_0) = f_1(E_1),$$
$$f_1^{-1}F_0 : E_0 \to E_1 \text{ is a vector bundle isomorphism } \xi_0 \to \xi_1,$$

and

$$\hat{F}(E_0 \times E_1) \text{ is open in } V \times I.$$

This last condition is automatic if $\partial M = \varnothing$.

One thinks of $\{F_t(E_0)\}_{t \in I}$ as a one-parameter family of tubular neighborhoods of M. Notice also that $\hat{F}$ defines a tubular neighborhood $(\hat{F}, \xi \times I)$ of $M \times I$ in $V \times I$.

It is easy to see that isotopy is an equivalence relation on the class of tubular neighborhoods for (V,M).

5.3. Theorem. *Let $M \subset V$ be a submanifold, $\partial M = \partial V = \varnothing$. Then any two tubular neighborhoods of M in V are isotopic.*

Proof. Let the tubular neighborhoods be $(f_i, \xi_i = (p_i, E_i, M))$, $i = 0, 1$. First suppose $f_0(E_0) \subset f_1(E_1)$.

Let $\Phi : \xi_0 \to \xi_1$ be the *fibre derivative* of $g = f_1^{-1} f_0 : E_0 \to E_1$. Thus Φ is the component along the fibres of the morphism

$$T_M g : T_M E_0 = TM \oplus \xi_0 \to TM \oplus \xi_1 = T_M E_1,$$

which shows that Φ is an isomorphism of vector bundles.

We define the *canonical homotopy from Φ to g* to be

$$H : E_0 \times I \to E_1,$$

$$H(x,y) = \begin{cases} t^{-1} g(tx) & \text{if } 1 \geqslant t > 0, \\ \Phi(x) & \text{if } t = 0. \end{cases} \tag{1}$$

(Here tx means scalar multiplication in the fibre containing x, etc.)

We claim H is C^∞. This is a local statement; to prove it we can work in charts for M, ξ_0 and ξ_1. Such charts make g locally a C^∞ embedding

$$g : U \times \mathbb{R}^k \to \mathbb{R}^m \times \mathbb{R}^k,$$
$$g(x,y) = (g_1(x,y), g_2(x,y)),$$
$$g(x,0) = (x,0)$$

where $U \subset \mathbb{R}^m$ is open. Locally, Φ becomes the fibre derivative of g:

$$\Phi : U \times \mathbb{R}^k \to \mathbb{R}^m \times \mathbb{R}^k,$$
$$\Phi(x,y) = \left(x, \frac{\partial g_2}{\partial y}(x,0) y \right).$$

Here $\partial g/\partial y$ assigns to each point of $U \times \mathbb{R}^k$ a linear map $\mathbb{R}^k \to \mathbb{R}^k$. The local representation of H is a map $(U \times \mathbb{R}^k) \times I \to \mathbb{R}^m \times \mathbb{R}^k$ given by the same formula (1).

By Taylor's formula we can write

$$g_2(x,y) = \frac{\partial g_2}{\partial y}(x,0)y + \langle s(x,y),y\rangle \tag{2}$$

where $\langle , \rangle$ is the inner product in $\mathbb{R}^k$ and $s: U \times \mathbb{R}^k \to \mathbb{R}^k$ is a C^∞ map with $s(x,0) = 0$.

It is trivial to verify that the first coordinate of (the local representation of) H as given by (1) defines a C^∞ map $U \times \mathbb{R}^m \times I \to \mathbb{R}^m$. By (2) the second coordinate of H is given by the formula

$$\frac{\partial g_2}{\partial y}(x,0)y + \langle s(x,ty),y\rangle, \qquad 1 \geqslant t \geqslant 0.$$

Clearly this is C^∞ in (t,x,y). Thus H is C^∞; and it is easily verified that H is an isotopy.

An isotopy of tubular neighborhoods from (f_0,ξ_0) to (f_1,ξ_1) is now defined by

$$F(x,t) = f_1^{-1}H(x,1 - t),$$

under the assumption that $f_0(E_0) \subset f_1(E_1)$.

For the general case we first pull E_0 into the open set $f_0^{-1}f_1(E_1) \subset F_0$ by a preliminary isotopy of the form

$$G: E_0 \times I \to E_0,$$
$$G(z,t) = (1 - t)y + th(y)$$

where

$$h: E_0 \to E_0,$$
$$h(y) = \left[\frac{\delta(p(y))}{1 + y^2}\right] y$$

and $\delta: M \to \mathbb{R}_+$ is a suitably small C^∞ map.

Thus (f_0,ξ_0) and (f_0G_1,ξ_0) are isotopic tubular neighborhoods; and since f_0G_1 maps E_0 into F_1, so are (f_0G_1,ξ_0) and (f_1,ξ_1). Theorem 5.3 now follows from transitivity of the relation of isotopy.

QED

6. Collars and Tubular Neighborhoods of Neat Submanifolds

The boundary of a manifold cannot have a tubular neighborhood. However, it has a kind of "half-tubular" neighborhood called a collar. A *collar on M* is an embedding

$$f: \partial M \times [0,\infty) \to M$$

such that $f(x,0) = x$. The following is the *collaring theorem*:

6.1. Theorem. *∂M has a collar.*

Proof. A proof using differential equations is given in Section 5.2. An alternative proof is outlined as follows.

First, find a C^∞ retraction $r: W \to \partial M$ of a neighborhood of ∂M onto ∂M. This is obviously possible locally, and two local retractions into coordinate domains can be glued together with a bump function. A standard globalization technique (e.g., Theorem 2.2.11) produces a global retraction.

Second, find a neighborhood $U \subset M$ of ∂M and a map

$$g: U \to [0,\infty),$$
$$g(\partial M) = 0$$

having 0 as a regular value. This is easily done with a partition of unity.

Third, observe that the map

$$h = (r,g): W \to \partial M \times [0,\infty)$$

maps a neighborhood of ∂M diffeomorphically onto a neighborhood $W \subset \partial M \times [0,\infty)$, and $h(x) = (x,0)$ for $x \in \partial M$.

Finally, let $\varphi: \partial M \times [0,\infty) \to h(W)$ be an embedding which fixes $\partial M \times 0$. Then $h^{-1}\varphi$ is a collar.

QED

It is also true that boundaries of C^0 manifolds have collars, although this is far from obvious. An elegant and surprising proof is given by M. Brown [2].

We leave as an exercise the proof of the following refinement of Theorem 6.1:

6.2. Theorem. *Let $M \subset V$ be a closed neat submanifold. Then ∂V has a collar which restricts to a collar on ∂M in M.*

Having collars at our disposal we can now prove:

6.3. Theorem. *Let $M \subset V$ be a neat submanifold. Then M has a tubular neighborhood in V.*

Proof. By Theorem 6.2 there is a neighborhood $N \subset V$ of ∂V and a diffeomorphism

$$\varphi: (N, \partial V) \approx (\partial V \times I, \partial V \times 0)$$

such that

$$\varphi: N \cap \partial M \approx \partial M \times I.$$

Let $q > 2 \dim V$. Embed ∂V in $\mathbb{R}^{q-1}$; extend this to an embedding

$$\partial V \times I \to \mathbb{R}^{q-1} \times [0,\infty) = \mathbb{R}^q_+$$
$$(x,t) \mapsto (x,t).$$

We can thus assume $N \subset \mathbb{R}^q_+$ in such a way that *every vector of $\mathbb{R}^q$ which is normal to N at a point of ∂V, or normal to $N \cap M$ at a point of ∂V, is in* $\mathbb{R}^{q-1}$. See Figure 4–2.

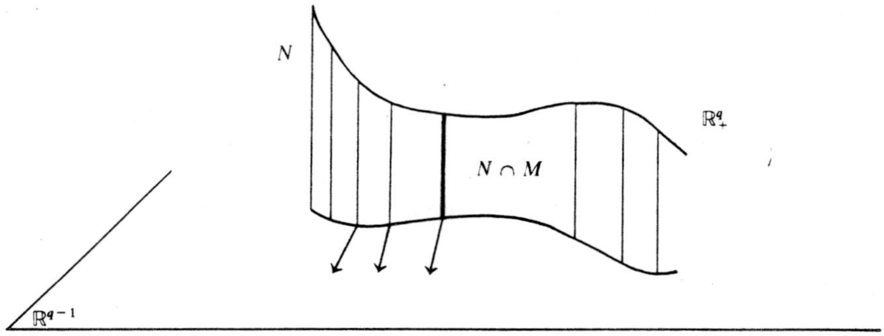

Figure 4–2. An embedding of N in $\mathbb{R}^q_+$.

We can extend the embedding of N to an embedding of V in $\mathbb{R}^q_+$. Thus V is now a neat submanifold of $\mathbb{R}^q_+$, and both V and M meet $\mathbb{R}^{q-1}$ orthogonally along ∂V and ∂M.

We can now find a normal tubular neighborhood of V in $\mathbb{R}^q_+$ (Figure 4–3), and the rest of the proof is like that of Theorem 5.2.

QED

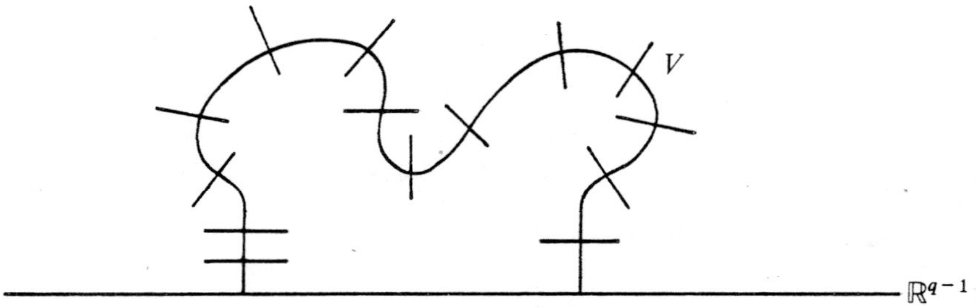

Figure 4–3. A normal tubular neighborhood of V in $\mathbb{R}^q_+$.

The following *extension theorem for tubular neighborhoods* is useful:

6.4. Theorem. *Let $M \subset V$ be a neat submanifold. Then every tubular neighborhood of ∂M in ∂V is the intersection with ∂V of a tubular neighborhood for M in V.*

Proof. First consider the special case $V = W \times I$, $M = N \times I$ where $N \subset W$ is a submanifold and $\partial N = \partial W = \varnothing$. Then

$$\partial V = W \times 0 \cup W \times 1$$
$$\partial N = N \times 0 \cup N \times 1.$$

In this case a tubular neighborhood for $(\partial V, \partial M)$ is just a pair of tubular neighborhood for (W,N). Let these be E_0, E_1. By Theorem 5.3 there is an isotopy of tubular neighborhoods from E_0 to E_1, say $F: E_0 \times I \to W$.

Then the corresponding embedding

$$\hat{F}: E_0 \times I \to W \times I = W$$
$$\hat{F}(x,t) = (F(x,t),t)$$

a tubular neighborhood for $N \times I = M$ in V, restricts to E_0 and E_1 in ∂V.

Now consider the general case. Give ∂V a collar in V which contain a collar on ∂M in M; we shall identify $\partial V \times [0,\infty)$ with a neighborhood of ∂V in V, so that $\partial M \times [0,\infty)$ corresponds to a neighborhood of ∂M in M. Put

$$V' = \partial V \times [0,1], \qquad M' = \partial M \times [0,1]$$
$$V'' = \partial V \times [1,\infty), \qquad M'' = \partial M \times [1,\infty).$$

Thus $V = V' \cup V''$, $V' \cap V'' = \partial V \times 1$, and similarly for M.

Let E_0 be a tubular neighborhood for ∂M in ∂V. By 6.3 there is a tubular neighborhood E'' of M'' in V''. Let $E_1 = E'' \cap \partial V \subset \partial V \times 1$. Thus E_0 and E_1 form a tubular neighborhood for $M \times \{0,1\}$ in $V \times \{0,1\}$. By the special case $E_0 \cup E_1$ extends to a tubular neighborhood E' of V' in M'. Then $E' \cup E''$ is a tubular neighborhood M in V which extends E_0. (Actually one has to make sure that E' and E'' fit together smoothly at $\partial V''$; this is left to the reader.)

QED

A *closed tubular neighborhood of radius* $\varepsilon > 0$, of a submanifold $M \subset V$, is an embedding $D_\varepsilon(\xi) \to M$ which is the restriction of a tubular neighborhood $(f,\xi = (p,E,M))$ of M. Here

$$D_\varepsilon(\xi) = \{x \in E : |x| \leqslant \varepsilon\}$$

is the *disk subbundle* of ξ of radius ε, for a given orthogonal structure on ξ.

The isotopy theorem for closed tubular neighborhoods is as follows:

6.5. Theorem. *Let $M \subset V$ be a submanifold. Let $\xi_i = (p_i,E_i,M)$ be orthogonal vector bundles over M, $i = 0, 1$. Let (f_i,ξ_i) be a tubular neighborhood of M. Let $\varepsilon > 0$, $\delta > 0$. Then (f_0,ξ_0) and (f_1,ξ_1) are isotopic by an isotopy of tubular neighborhoods $F_t: E \to V$, $0 \leqslant t \leqslant 1$, such that $F_0 = f_0$ and*

$$F_1(D_\varepsilon(\xi_0)) = D_\delta(\xi_1).$$

Proof. By Theorem 5.3 and a preliminary isotopy we may assume that, as tubular neighborhoods, $(f_0,\xi_0) = (f_1,\xi_1)$; but $D_\varepsilon(\xi_0)$ and $D_\delta(\xi_1)$ might be defined by different orthogonal structures. However, by 2.3 and a linear isotopy we may assume that these orthogonal structures are identical. The theorem is now obvious: in any orthogonal vector bundle there is a linear isotopy carrying $D_\varepsilon(\xi)$ onto $D_\delta(\xi)$.

QED

As a very special but useful case, let M be a point $x_0 \in V$. An open tubular neighborhood of x_0 is an embedding $(\mathbb{R}^n,0) \to (V,x_0)$ and a closed tubular neighborhood of radius 1 is an embedding $(D^n,0) \to (V,x_0)$, where $n = \dim V$. (We suppose $\partial V = \varnothing$.)

6.6. Theorem. *Let* $E^n = D^n$ *or* $\mathbb{R}^n$ *and let* $f_i:(E^n,0) \to (V,x_0)$ *be an embedding,* $i = 0, 1$, *where* $n = \dim V$. *If*

$$\mathrm{Det}(D(f_1^{-1}f_0)(0)) > 0$$

then f_0 *and* f_1 *are isotopic* rel 0.

Proof. Note that $f_1^{-1}f_0$ is well defined on a neighborhood of 0 in E^n, so its derivative at 0 is defined. By isotopy of tubular neighborhoods we can assume $f_1^{-1}f_0$ is a linear automorphism $L \in GL(n)$. If $\det L > 0$ then L is connected to the identity in $GL(n)$ by an arc L_t, $0 \leqslant t \leqslant 1$:

$$L_0 = f_1^{-1}f_0, \qquad L_1 = 1_{\mathbb{R}^n}.$$

The required isotopy from f_0 to f_1 is

$$f_1 L_t^{-1} f_1^{-1} f_0, \qquad 0 \leqslant t \leqslant 1.$$

QED

One use for tubular neighborhoods is to make a map look like a vector bundle map (after a homotopy). Let V, N be manifolds, $A \subset N$ a compact neat submanifold and $f:V \to N$ a map such that f and $f|\partial V$ are both transverse to A. Put $M = f^{-1}(A)$, a neat submanifold of V. Suppose given tubular neighborhoods $U \subset V$ of M and $E \subset N$ of A. Let $D \subset U$ be a disk subbundle such that $f(D) \subset E$.

6.7. Theorem. *Under the assumptions above, there is a homotopy* f_t *from* $f = f_1$ *to a map* $f_0 = h:V \to N$ *such that*:

(a) $h|D$ *is the restriction of a vector bundle map* $U \to E$ *over* $f:M \to A$;
(b) $f_t = f$ on $M \cup (N - U)$, $0 \leqslant t \leqslant 1$.
(c) $f_t^{-1}(N - A) = V - M$, $0 \leqslant t \leqslant 1$.

Proof. Let $\Phi:U \to E$ be the vector bundle map, over $f:M \to A$, which is the fibre derivative of $f:D \to E$. Let $f_t:D \to E$, $1 \geqslant t \geqslant 0$ be the canonical homotopy from $f_1 = f|D$ to $f_0 = \Phi|D$:

$$f_t(x) = \begin{cases} t^{-1}f(tx), & 1 \geqslant t \geqslant 0, \\ \Phi(x), & t = 0. \end{cases}$$

Notice that $f_t(\partial D) \subset N - A$.

Let $D' \subset U$ be a disk subbundle such that D' int D. Put $D' - \operatorname{int} D = L$; thus $\partial L = \partial D' \cup \partial D$. Define a homotopy

$$g_t:\partial L \to N - A,$$

$$g_t = \begin{cases} f_t & \text{on} \quad \partial D, \\ f & \text{on} \quad \partial D. \end{cases}$$

By homotopy extension (Theorem 1.4) g_t extends to a homotopy

$g_t: L \to N - A$. Define a homotopy

$$h_t: V \to N,$$

$$h_t = \begin{cases} f & \text{on} \quad V - D \\ g_t & \text{on} \quad L \\ f_t & \text{on} \quad D. \end{cases}$$

The required map is then $h = h_0$.

QED

Exercises

1. There is an obvious definition of C^r tubular neighborhood and C^r isotopy, $1 \leqslant r < \infty$. A neat C^r submanifold has a C^r tubular neighborhood, unique up to C^r isotopy.

***2.** Let M_0, M_1 be neat submanifolds of V in general position. Let (f_i,ξ_i) be a tubular neighborhood of $M_0 \cap M_1$ in M_i, $i = 0, 1$. Then there is a tubular neighborhood $(f,\xi_0 \oplus \xi_1)$ of $M_0 \cap M_1$ in V such that $f|\xi_i = f_i$.

***3.** *Extendability of germs of tubular neighborhoods.* Let $M \subset V$ be a neat submanifold and $U \subset M$ a neighborhood of a closed subset $A \subset M$. For every tubular neighborhood E_0 of U in V there is a tubular neighborhood E of M in V, and a neighborhood $W \subset U$ of A, such that $E|W = E_0 W$.

4. Let $D \subset M$ be a neat p-disk of codimension k. Then K has a neighborhood $E \subset M$ such that

$$(E,D) \approx (D^p \times \mathbb{R}^k, D^p \times 0).$$

5. Let $M \subset V$ be a closed neat submanifold of codimension k. Then there is a map $f:(V,M) \to (S^k,p)$ such that p is a regular value and $f^{-1}(p) = M$, if and only if M has a trivial normal bundle.

6. Let $M \subset \mathbb{R}^n$ be a submanifold of codimension k, $\partial M = \varnothing$. Let $v: M \to G_{n,k}$ be a transverse field of k-planes. Suppose that v locally satisfies Lipschitz conditions with respect to Riemannian metrics on M and $G_{n,k}$. Then M has a tubular neighborhood $U \subset \mathbb{R}^n$ whose fibre over $x \in M$ is the intersection of U and the k-plane through x parallel to $v(x)$. But if v is merely continuous this may be false, even for $S^1 \subset \mathbb{R}^2$.

7. The boundary of a nonparacompact manifold does not necessarily have a collar. For instance there is a 2-dimensional manifold M such that $M - \partial M \approx \mathbb{R}^2$ but ∂M has uncountably many components (each diffeomorphic to $\mathbb{R}$).

*****8.** Let L be the long line with its natural ordering (see Exercise 2, Section 1.1) and set

$$M = \{(x,y) \in L \times L : x \leqslant y\}.$$

Then M is a ∂-manifold with $\partial M \approx L$. It seems unlikely that ∂M has a collar.

***9.** *Ambient isotopy of closed tubular neighborhoods.* In Theorem 6.5, the isotopy $F_t|D_\varepsilon(\xi_0)$ can be achieved through a diffeotopy of V. That is, there is an isotopy $G_t: V \to V$, $0 \leqslant t \leqslant 1$, such that $G_0 = 1_V$, and $F_t(x) = G_t f_0(x)$ for $x \in D_\varepsilon(\xi_0)$.

7. Analytic Differential Structures

We shall use tubular neighborhoods and transversality to prove the following result:

7.1. Theorem (Whitney). *Let M be a compact manifold without boundary. Then M is diffeomorphic to an analytic submanifold of Euclidean space.*

Proof. We may assume M embedded in $\mathbb{R}^q$ with codimension k. Let $E \subset \mathbb{R}^q$ be a normal tubular neighborhood of M. We identify E with a neighborhood of the zero section of the normal bundle of M. Let $p:E \to M$ be the restriction of the bundle projection.

Let $g:M \to G_{q,k}$ be the map sending $x \in M$ to the k-plane normal to M at x. Let $E_{q,k} \to G_{q,k}$ be the Grassmann k-plane bundle and let $f:E \to E_{q,k}$ be the natural map covering h; thus

$$f(y) = (h(y),y) \in E_{q,k} \subset G_{q,k} \times \mathbb{R}^q.$$

Note that f is transverse to the zero section $G_{q,k} \subset E_{q,k}$ and

$$f^{-1}(G_{q,k}) = M.$$

The main point of the proof is to C^1 approximate f by an analytic map. For this we use Theorem 2.5.2. That result says that a real-valued C^r map $(0 \leqslant r \leqslant \infty)$ on an open subset of Euclidean space can be C^r approximated, near a compact set, by an analytic map. The same result clearly holds for maps into $\mathbb{R}^s$. Moreover it holds for maps from open subsets $E \subset \mathbb{R}^q$ into a C^ω submanifold $N \subset \mathbb{R}^s$. For a normal tubular neighborhood of N provides a C^ω retraction $\rho: W \to N$ where $W \subset \mathbb{R}^q$ is an open set. Given $f:E \to N$, the required C^ω approximation is $\rho \circ f'$ where $f':E \to W$ is a C^ω approximation to f.

Now $E_{q,k}$ embeds analytically in $\mathbb{R}^s$ with $s = q^2 + q$. For this it suffices to embed $G_{q,k}$ in $\mathbb{R}^{q^2}$. This is done by mapping a k-plane $P \in G_{q,k}$ to the linear map $\mathbb{R}^q \to \mathbb{R}^q$ given by orthogonal projection on P.

It follows that the map $f:E \to E_{q,k}$ can be approximated near M by an analytic map $\varphi:E \to E_{q,k}$. Put $M' = \varphi^{-1}(G_{q,k})$. If φ is sufficiently C^1 close to f then $\varphi \pitchfork G_{q,k}$ and the restriction of $p:E \to M$ to M' is a C^∞ diffeomorphism $M' \approx M$.

QED

Of course stronger results can be proved by using the powerful Remmert–Grauert approximation Theorem 2.5.1. The proof given used only the elementary Theorem 2.5.2.

Exercises

1. Let $f:M \to \mathbb{R}^q$ be an embedding where M is compact without boundary. Then f can be approximated by embeddings g such that $g(M)$ is an analytic submanifold.

2. Let $M \subset \mathbb{R}^s$ and $N \subset \mathbb{R}^t$ be analytic submanifold without boundaries, with M compact. Then analytic maps are dense in $C^r_W(M,N)$, $0 \leqslant r \leqslant \infty$.

3. Let ξ be a C^r vector bundle over M, where M is a compact analytic submanifold of Euclidean space, $0 \leqslant r \leqslant \infty$, and $\partial M = \varnothing$. Then ξ has a compatible C^ω vector bundle structure, unique up to C^ω isomorphism.

Chapter 5

Degrees, Intersection Numbers, and the Euler Characteristic

Topology has the peculiarity that questions belonging in its domain may under certain circumstances be decidable even though the continua to which they are addressed may not be given exactly, but only vaguely, as is always the case in reality.

—H. Weyl, *Philosophy of Mathematics and Natural Science*, 1949

Geometry is a magic that works . . .

—R. Thom, *Stabilité Structurelle et Morphogénèse*, 1972

We now have enough machinery at our disposal to develop one of the most important tools in topology: the degree of a map $f: M \to N$, where M and N are compact n-manifolds, N is connected, and $\partial M = \partial N = \varnothing$. This degree is an integer if M and N are oriented, an integer mod 2 otherwise.

Intuitively, the degree is the number of times f wraps M around N. The precise definition requires the theories of approximation, regular values, and orientation. If f is C^1 and if $y \in N$ is a regular value, then the degree of f is the number of points in $f^{-1}(y)$ at which Tf preserves orientation, minus the number of points at which Tf reverses orientation.

It turns out that the degree of f is the same for all maps homotopic to f. This has two important consequences: it makes the degree of any given map easy to compute, and it gives us a convenient method of distinguishing homotopy classes. Moreover the degree is the *only* homotopy invariant for maps into S^n; this is the main result of Section 5.1.

With the introduction of the degree we enter the realm of algebraic topology. Many geometrical questions depend on the computation of degrees of maps; thus topology is translated to algebra, the continuous is reduced to the discrete.

The degree is actually a special case of a more general geometrical concept called the *intersection number*, developed in Section 5.2. If M and N are submanifolds of W of complementary dimensions, and M and N are in general position, their intersection number is the algebraic number of points in $M \cap N$, each counted with appropriate sign determined by orientations. By means of transversality theory, intersection numbers of maps $M, N \to W$ can be defined; again we obtain homotopy invariants. If W is an n-dimensional oriented vector bundle $\xi \to M$ then the self-intersection number of the zero section is called the *Euler number* $X(\xi)$. This is an important isomorphism invariant of bundles. The Euler number of TM is the *Euler characteristic* $\chi(M)$.

We can compute $X(\xi)$ by means of sections of ξ. This leads to the computation of $\chi(M)$ as the sum of the indices of zeros of a vector field on M. In Chapter 6 we shall use the Morse inequalities to recompute $\chi(M)$ as the alternating sum of the Betti numbers of M.

1. Degrees of Maps

In this section we exploit orientations and tubular neighborhoods to derive some classical homotopy and extension theorems.

Recall that Euclidean n-space $\mathbb{R}^n$, $n \geqslant 1$, has the *standard orientation* ρ^n given by any basis whose coordinate matrix has positive determinant (and the orientation of $\mathbb{R}^0$ is the number $+1$). Every n-dimensional submanifold of $\mathbb{R}^n$ is also given this orientation.

If (M,ω), (N,θ) are oriented n-manifolds, the *product orientation* $\omega \times \theta$ for $M \times N$ assigns to $(x,y) \in M \times N$ the orientation $\omega_x \oplus \theta_y$ of $(M \times N)_{(x,y)} = M_x \oplus N_y$.

Let (M,ω) be an oriented ∂-manifold and $\partial f: \partial M \times [0,\infty) \to M$ a collar. Then $T_{\partial M} f$ induces an isomorphism of the trivial bundle $M \times \mathbb{R}$ onto the normal bundle ν of ∂M in M. The standard orientation of $\mathbb{R}$ orients each fibre of $M \times \mathbb{R}$; via $T_{\partial M} f$ this induces an orientation ι of ν which does not depend on the collar. In other words ν is oriented by *inward* pointing vectors tangent to M at ∂M.

We now have orientations ω and i of M and ν. From the exact sequence of vector bundles

$$0 \to T(\partial M) \to T_{\partial M} M \to \nu \to 0$$

we define the *induced orientation* $\omega/\iota = \partial\omega$ of ∂M. Thus $(e_1, \ldots, e_{n-1})$ is an orienting basis for $(\partial\omega)_x$ if $(e_1, \ldots, e_{n-1}, e_n)$ is an orienting basis for ω_x and e_n points *into* M at $x \in \partial M$.

Let θ be an orientation of M. We usually give $M \times I$ the product orientation $\omega = \theta \times \rho^1$ where ρ^1 is the standard orientation of I. It follows that

$$\partial\omega | M \times 0 = \theta \qquad \text{and} \qquad \partial\omega | M \times 1 = -\theta.$$

We shall frequently speak of "the oriented manifold M", not naming the orientation explicitly. In this case ∂M and $M \times I$ are also oriented manifolds, as is any submanifold of M of the same dimension. If $-M$ denotes the manifold M with the opposite orientation, then

$$\partial(M \times I) = (M \times 0) \cup (-M \times 1)$$

as oriented manifolds.

The closed unit n-disk $D^{n+1} \subset \mathbb{R}^{n+1}$ has the standard orientation. Therefore its boundary S^n inherits an orientation, also called "standard". It is easy to verify that stereographic projection from the north pole $P = (0, \ldots, 0,1) \in S^n$ is an orientation preserving diffeomorphism $S^n - P \approx \mathbb{R}^n$. Thus if $(e_1, \ldots, e_n)$ is an orienting basis for $\mathbb{R}^n \subset \mathbb{R}^{n+1}$, an orienting basis for S^n at the south pole $-P$ is $(e_1, \ldots, e_n)$, while at the north pole $(e_1, \ldots, e_{n-1}, -e_n)$ is orienting.

Let $A: \mathbb{R}^m \to \mathbb{R}^m$ be the antipodal map $A(x) = -x$. Since $\operatorname{Det} A = (-1)^m$ it follows that A preserves orientation of $\mathbb{R}^m$ if and only if m is even. The antipodal map of $\mathbb{R}^{n+1}$ restricts to a diffeomorphism of D^{n+1}. Since it clearly preserves orientation of the normal bundle of ∂D^{n+1}, it follows that $A: S^n \to S^n$ preserves orientation if and only if n is odd.

1.1. Lemma. *Let (W,ω) be an oriented ∂-manifold. Suppose $K \subset W$ is an embedded arc which is transverse to ∂W at its endpoints $u, v \in \partial W$. Let κ be an orientation of K, and consider the quotient orientation ω/κ of the algebraic normal bundle of κ. Then*

$$\omega_u/\kappa_u = (\partial\omega)_u \Leftrightarrow \omega_v/\kappa_v = -(\partial\omega)_v.$$

Proof. Let X_u, X_v be tangent vectors to K at u, v which belong to κ_u, κ_v respectively. Then X_u is inward if and only if X_v is outward; this is equivalent to the lemma.

QED

Let (M,ω), (N,θ) be compact oriented manifolds of the same dimension, without boundaries. Assume N is connected. Let $f: M \to N$ be a C^1 map and $x \in M$ a regular point of f. Put $y = f(x)$. We say x has *positive type* if the isomorphism $T_x f: M_x \to N_y$ preserves orientation, that is, it sends ω_x to θ_y. In this case we write $\deg_x f = 1$. If $T_x f$ reverses orientation then x has *negative type*, and we write $\deg_x f = -1$. We call $\deg_x f$ the *degree of f at x*.

Suppose $y \in N$ is any regular value for f. Define the *degree of f over y* to be

$$\deg(f, y) = \textstyle\sum_{x \in f^{-1}(y)} \deg_x f;$$

if $f^{-1}(y)$ is empty, $\deg(f,y) = 0$. To indicate orientations we also write

$$\deg(f, y) = \deg(f, y; \omega, \theta).$$

Reversing ω or θ changes the sign of $\deg(f, y)$.

To interpret $\deg(f, y)$ geometrically, suppose that $f^{-1}(y)$ contains n points of positive type and m points of negative type, so that $\deg(f, y) = n - m$. From the inverse function theorem we can find an open set $U \subset N$ about y and an open set $U(x) \subset M$ about each $x \in f^{-1}(y)$ such that f maps each $U(x)$ diffeomorphically onto U preserving or reversing orientation according to the type of x. Thus $\deg(f, y)$ is the algebraic number of times f covers U.

For example, let S^1 be the unit circle in the complex plane. Let $M = N = S^1$, and $\theta = \omega$. If $f: S^1 \to S^1$ is the map $f(z) = z^n$ then $\deg(f,z) = n$, provided $z \neq 1$ when $n = 0$.

If M is not connected, but has components $M_1, \ldots, M_k$, note that

$$\deg f = \textstyle\sum_i \deg(f|M_i).$$

Of course each M_j is given the orientation $\omega|M_j$ induced by the inclusion $M_j \hookrightarrow M$.

1.2. Lemma. *Let W be a compact oriented manifold of dimension $n + 1$, N a compact oriented n-manifold without boundary and $h: W \to N$ a C^∞ map. Let $y \in N$ be a regular value for both h and $h|\partial W$. Then* $\deg(f,y) = 0$.

Proof. Let ω, θ be the orientations of W, N respectively. Let $M_1, \ldots, M_m$ be the components of ∂W.

Since y is a regular value, $h^{-1}(y)$ is a compact 1-dimension submanifold of W whose boundary is $(h|\partial W)^{-1}(y)$. Let $u \in h^{-1}(y)$. Then there is a unique $v \in h^{-1}(y)$, $v \neq u$, and a component arc $K \subset h^{-1}(y)$ such that $\partial K = \{u,v\}$. It suffices to show that u and v are of opposite type for $h|\partial W$.

Let $\nu = TW/TK$, the algebraic normal bundle of K in W. Since y is a regular value, Tf induces a bimorphism $\Phi: \nu \to N_y$. There are natural identifications $\nu_u = (\partial W)_u$, $\nu_v = (\partial W)_v$.

Since K is an arc there is a unique orientation κ of ν such that $\kappa_u = (\partial\omega)_u$. By Lemma 1.1, $\kappa_v = -(\partial\omega)_v$.

Suppose u is of positive type (for $h|\partial W$). Then $\Phi_x(\kappa_x) = \theta_y$ for all $x \in K$. It follows that

$$\begin{aligned} T(h|\partial W)(\partial\omega)_u &= \Phi_u \kappa_u \\ &= \theta_y \end{aligned}$$

and

$$\begin{aligned} T(h|\partial W)(-\partial\omega)_v &= \Phi_v \kappa_v \\ &= \theta_v. \end{aligned}$$

Therefore v is of negative type. This shows that $h|\partial W$ has equal numbers of points of positive and negative type in $h^{-1}(y)$.

QED

1.3. Corollary. *Let (M,ω) and (N,θ) be compact, oriented n-manifolds, $\partial M = \partial N = \varnothing$. Let N be connected, and $f, g: M \to N$ homotopic C^∞ maps having a common regular value $y \in N$. Then* $\deg(f,y) = \deg(g,y)$.

Proof. There is a homotopy $h: M \times I \to N$ from f to g, and one can make h C^∞ and transverse to y. As oriented manifolds

$$\partial(M \times I) = (M \times 0, \omega) \cup (M \times 1, -\omega).$$

By Theorem 1.2 we have

$$\begin{aligned} 0 &= \deg(h|\partial(M \times I)) \\ &= \deg(f,y;\, \omega,\theta) + \deg(g,y;\, -\omega,\theta) \\ &= \deg(f,y;\, \omega,\theta) - \deg(g,y;\, \omega,\theta). \end{aligned}$$

QED

1.4. Lemma. *Let M, N be a compact oriented n-manifolds without boundaries, $n \geqslant 1$, with N connected. Let y, $z \in N$ be regular values for a C^∞ map $f:M \to N$. Then* $\deg(f,y) = \deg(f,z)$.

Proof. Suppose there is a diffeomorphism $h:N \to N$, homotopic to the identity, such that $h(y) = z$. Then $\deg(h,z) = \deg_y h = 1$, by Theorem 1.3. It is easy to see that this implies

$$\deg(f,y) = \deg(hf,z).$$

But hf is homotopic to f; hence

$$\deg(hf,z) = \deg(h,z).$$

It remains to construct h. If y and z are very close together, say in the same coordinate ball, the construction is not hard, and is left as an exercise. The relation between y and z, that such an h exists, is an equivalence relation on N whose equivalence classes are thus disjoint open sets. Since N is connected, any two points are equivalent.

QED

1.5. Lemma. *Let M, N be manifolds and $f:M \to N$ a continuous map. Then f can be approximated by C^∞ maps homotopic to f.*

Proof. We may assume, by Theorem 4.6.3, that N is a C^∞ retract of an open subset $W \subset \mathbb{R}^n_+$; let $r:W \to N$ be a retraction. Let $g:M \to N$ be a C^∞ map which approximates f so closely that the map

$$h:M \times I \to \mathbb{R}^n_+,$$
$$h(x,t) = (1 - t)f(x) + tg(x)$$

takes value in W. Then $rh:M \times I \to N$ is a homotopy from f to g.

QED

We are ready to define the *degree of a map*. Let M, N be oriented compact n-manifolds, $n \geqslant 1$, with N connected and $\partial M = \partial N = \varnothing$. The *degree* $\deg f$ of a continuous map $f:M \to N$ is defined to be $\deg(g,z)$ where $g:M \to N$ is a C^∞ map homotopic to f and $z \in N$ is a regular value for g. By Theorem 1.5 such a g exists, and $\deg f$ is independent of g and z by Theorems 1.3 and 1.4.

If M and N are not oriented, perhaps even nonorientable, a *mod* 2 *degree* of $f:M \to N$ is defined as follows. Again let $z \in N$ be a regular value for a C^∞ map $g:M \to N$ homotopic to f. Let $\deg_2(g,z)$ denote the reduction modulo 2 of the number of points in $g^{-1}(z)$. Then $\deg_2(g,z)$ is independent of g and z. This follows from the mod 2 analogue of Lemma 1.2, the proof of which reduces to the fact that a compact 1-manifold has an even number of boundary points. We then define $\deg_2(f) = \deg_2(g,z)$.

The results proved up to now apply to degrees of continuous maps to yield:

1.6. Theorem. *Let M, N be compact n-manifolds without boundary, with N connected.*

(a) *Homotopic maps $M \to N$ have the same degree if M, N are oriented, and the same mod 2 degree otherwise.*

(b) *Let $M = \partial W$, W compact. Suppose a map $f: M \to N$ extends to W. Then $\deg f = 0$ if W and N are orientable, and $\deg_2 f = 0$ otherwise.*

The degree is a powerful tool in studying maps. For example, *if* $\deg f$ *(or $\deg_2 f$) is nonzero then f must be surjective.* For if f is not surjective, it can be approximated by a homotopic C^∞ map g which is not surjective. If $y \in N - g(M)$, clearly $\deg(g,y) = 0$.

Here is an application of degree theory to complex analysis; it has the fundamental theorem of algebra as a corollary. Let $p(z)$, $q(z)$ be complex polynomials. The rational function $p(z)/q(z)$ extends to a C^∞ map $f: S^2 \to S^2$, where S^2 denotes the Riemann sphere (the compactification of the complex field $\mathbb{C}$ by ∞). Then: *f is either constant or surjective.*

The key to the proof is the observation that $z \in S^2$ is a regular point if and only if the complex derivative $f'(x) \neq 0$, and in this case the real derivative $Df_z: \mathbb{R}^2 \to \mathbb{R}^2$ has *positive* determinant.

If f is not constant then f' is not identically 0; hence there is a regular point z. By the inverse function theorem there is an open set $U \subset S^2$ about z, containing only regular points, such that $f(U)$ is open. Let $w \in f(U)$ be a regular value. Then $f^{-1}(w)$ is nonempty. Since every point in $f^{-1}(w)$ has positive type, it follows that $\deg(f,w) = \deg f > 0$. Therefore f is surjective.

A famous application of degree theory is the so-called "hairy ball theorem": *every vector field on S^{2n} is zero somewhere*; more picturesquely, a hairy ball cannot be combed. To prove this, suppose that σ is a vector field on S^k which is nowhere zero. A homotopy of S^k from the identity to the antipodal map is obtained by moving each $x \in S^k$ to $-x$ along the great semicircle in the direction $\sigma(x)$. The existence of such a homotopy implies that the antipodal map has degree $+1$ and so preserves orientation; therefore k is odd.

The question of zeros of a vector field, or more generally, of a section of a vector bundle, is approached more systematically in Section 5.2 with the theory of Euler numbers.

The following lemma will be used in the extension Theorem 1.8.

1.7. Lemma. *Let W be an oriented $(n+1)$-manifold and $K \subset W$ a neat arc. Let $V \subset \partial W$ be a neighborhood of ∂K and $f: V \to N$ a map to an oriented manifold N, $\partial N = \varnothing$. Let $y \in N$ be a regular value of f and assume $\partial K = f^{-1}(y)$. Finally, assume that f has opposite degrees at the two endpoints of K. Then there is a neighborhood $W_0 \subset W$ of K and a map $g: W_0 \to N$ such that:*

(a) *$g = f$ on $W_0 \cap V$,*

(b) *y is a regular value of g,*

(c) *$g^{-1}(y) = K$.*

Proof. We may take $(N,y) = (\mathbb{R}^n,0)$. Let the endpoints of K be x_0, x_1. Since 0 is a regular value each x_i has a neighborhood $U_i \subset V$ such that f restricts to an embedding $f_i:(U,x_i) \to (\mathbb{R}^n,0)$.

It suffices to prove the lemma for any map agreeing with f near K. Therefore we can assume that each f_i is a diffeomorphism. Then f_i^{-1} can be regarded as a tubular neighborhood of x_i in W; and together, f_0^{-1} and f_1^{-1} form a tubular neighborhood of ∂K in ∂W.

By Theorem 4.6.4 this tubular neighborhood extends to a tubular neighborhood E of K in V. We may assume $W = E$. Since K is an arc, E is a trivial vector bundle over K, and we may assume that

$$(W,K) = (I \times \mathbb{R}^n, I \times 0)$$

and $(N,y) = (\mathbb{R}^n,0)$. With this notation,

$$V = 0 \times \mathbb{R}^n \cup 1 \times \mathbb{R}^n$$

and $f_i:i \times \mathbb{R}^n \to \mathbb{R}^n$ $(i = 0, 1)$ is given by a linear isomorphism $L_i \in GL(n)$. The degree assumptions and the convention for orienting $\partial(I \times \mathbb{R}^n)$ mean that L_0 and L_1 have determinants of the *same* sign. Therefore L_0 and L_1 can be joined by a path L_t in $GL(n)$, $0 \leqslant t \leqslant 1$. The required extension of f is the map

$$I \times \mathbb{R}^n \to \mathbb{R}^n,$$
$$(t,y) \to L_t(y).$$

QED

We can now prove a basic extension theorem:

1.8. Theorem. *Let W be a connected oriented compact ∂-manifold of dimension $n + 1$. Let $f:\partial W \to S^n$ be a continuous map. Then f extends to a map $W \to S^n$ if and only if* $\deg f = 0$.

Proof. We already know that the degree vanishes if f extends. Suppose then that $\deg f = 0$.

By homotopy extension it suffices to extend some map homotopic to f. Since f is homotopic to a C^∞ map (Theorem 1.5) we may assume f is C^∞. Let $y \in S^n$ be a regular value of f.

Since $\deg(f,g) = 0$, $f^{-1}(y)$ has equal numbers of points of positive and negative type. We can find a set of disjoint embedded arcs $K_1, \ldots, K_m \subset W$, each going from a positive to a negative point of $f^{-1}(y)$, with $K = K_1 \cup \cdots \cup K_m$ a neat submanifold and $\partial K = f^{-1}(y)$. When $\dim W \geqslant 3$ this follows from density of embeddings $I \to W$. When $\dim W = 2$ we can find immersed arcs $K_1, \ldots, K_m$ (which may across each other). A new family of $K_1', \ldots, K_m'$ without crossings can be obtained by the following device. Assume the crossings are in general position. At each crossing make the change suggested by Figure 5–1. The arrows indicate the orientation of the arcs from positive to negative endpoint. There results a compact neat 1-dimensional submanifold K' of W with boundary $f^{-1}(y)$. Each component

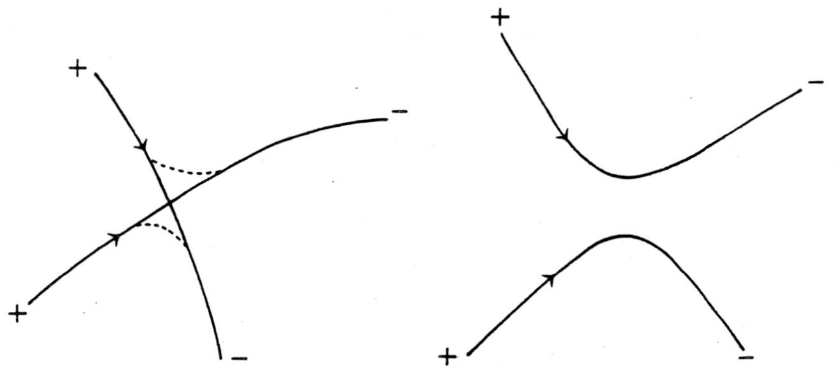

Figure 5–1. Eliminating crossings.

of K' is an arc or a circle; each arc has endpoints of opposite types (see the arrows in Figure 5–1). Thus we obtain disjoint embedded arcs.

Now apply Lemma 1.7 to each arc K_i, with $N = S^n$. We obtain an open neighborhood $W_0 \subset W$ of $\cup K_i$ and a map $g: W_0 \to S^n$ which agrees with f on ∂W_0, having y as a regular value, and with $g^{-1}(y) = K_i$.

Let $U \subset W_0$ be a smaller open neighborhood of K_i whose closure is in W_0. Then Bd $U \subset W_0 - \cup K_i$. The maps g and f fit together to form a continuous map

$$H: X = \text{Bd } U \cup (\partial W - U) \to S^n - y.$$

Note that X is a closed subset of $W - U$. Since $S^n - y \approx \mathbb{R}^n$, Tietze's extension theorem permits an extension of h to a map $H: W - U \to S^n - y$. An extension of f to W is the map equal to H on $W - U$ and to g on W_0.

QED

An analogue of Theorem 1.8 for nonorientable manifolds is:

1.9. Theorem. *Let W be a connected compact nonorientable ∂-manifold of dimension $n + 1 \geqslant 2$. A map $f: \partial W \to S^n$ extends to W if and only if* $\deg_2 f = 0$.

Proof. If f extends, $\deg_2 f = 0$ by Theorem 1.6.

Suppose $\deg_2 f = 0$. We may assume f is C^∞. Consider first the case $\dim W \geqslant 3$.

Let $y \in S^n$ be a regular value; then $f^{-1}(y)$ has even cardinality. Hence $f^{-1}(y) = \partial K$ where $K \subset W$ is the union of disjoint neatly embedded arcs.

Let K_i be one of these arcs with endpoints $u, v \in f^{-1}(y)$. Although TW is not an orientable vector bundle, $TW|K_i$ is. Give $TW|K_i$ an arbitrary orientation; this induces orientations to $T_u\partial W$ and $T_v\partial W$. It then makes sense to ask whether u and v are of opposite type for the map f. If they are not, we change K_i to a new arc K_i' by adding to it an orientation reversing loop L in W (see Figure 5–2).

Because dim $W \geqslant 3$ we can make K_1' embedded and disjoint from the other arcs K_j. Give $TW|K_i'$ an orientation and give $T_u\partial W$ and $T_v\partial W$ the induced orientations. With respect to these orientations, u and v are now of opposite type; otherwise L would preserve orientation.

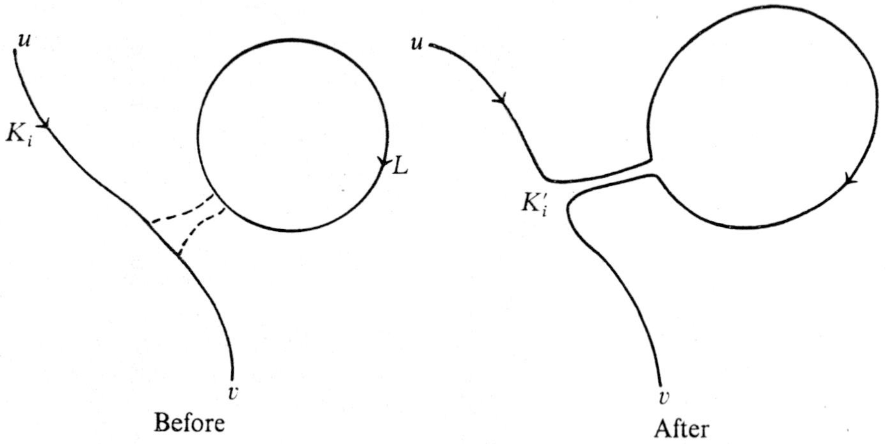

Figure 5–2.

We can thus assume that T_KW is oriented so that with respect to the induced orientation of $T(\partial W)|f^{-1}(y)$, the endpoints of each arc are of opposite type. The rest of the proof for dim $W \geqslant 3$ is now exactly like that for the oriented case.

Now let dim $W = 2$. Let $y \in S^1$ be a regular value for f. After a homotopy, we may assume that there are disjoint open intervals $I_1, \ldots, I_v \subset \partial W$ with the following properties:

(a) each I_j contains exactly one point x_i of $f^{-1}(y)$;
(b) f maps I_j diffeomorphically onto $S^1 - (-y)$;
(c) $f(\partial W - \cup I_j) = -y$.

We say that f is in *standard form* in this case.

Give ∂W any orientation, so that the integer deg f is now defined. Note that deg f is even. Each I_j contributes ± 1 to deg f; hence v is even.

We proceed by induction on $v = v(f)$; if $v = 0$ then f is constant and extends to the constant map of W. Suppose then that $v \geqslant 2$.

Let $K \subset W$ be a neat arc joining x_1 to x_2; give T_KW an orientation ω. As before, choose K so that x_1 and x_2 are of opposite type for f with respect to the orientations of $T_{x_i}\partial W$, $i = 1, 2$ induced by ω.

Let $N \subset W$ be a tubular neighborhood of K such that $N \cap \partial W = I_1 \cup I_2$. Topologically N is a rectangle whose boundary ∂N is a circle consisting of four arcs I_1, J_1, I_2, J_2.

We change f to a new map $g: \partial W \to S^1$ thus: $g = f$ on $\partial W - (I_1 \cup I_2)$, while $g(I_1 \cup I_2) = -y$. Observe that g is in standard form and $\nu(g) = \nu(f) - 2$. We assume inductively that g extends to a map $G: W \to S^1$.

Note that $\deg(G|\partial N) = 0$, by Theorem 1.6(b) since G extends over N. Since $G|I_1 \cup I_2$ is constant, this means that $\deg G|J_1 + \deg G/J_2 = 0$, when J_1 and J_2 are given orientations induced from an orientation of ∂N.

Define a new map $h: \partial N \to S^1$ equal to G on $J_1 \cup J_2$ and to f on $I_1 \cup I_2$. Because of the way the arc K was chosen, $\deg f|I_1 + \deg f|I_2 = 0$. It follows that $\deg h = 0$. By Theorem 1.8 (adapted to N) there is an extension of h to a map $H: N \to S^1$. The required extension of f is the map $W \to S^1$ which equals H on N and G on $W - N$.

QED

We can now classify maps of all compact n-manifolds into S^n. Let $\simeq$ denote the relation of homotopy.

1.10. Theorem. *Let M be a compact connected n-manifold, $n \geqslant 1$. Let $f, g: M \to S^n$ be continuous maps.*

(a) *If M is oriented and $\partial M = \varnothing$, then $f \simeq g$ if and only if* $\deg f = \deg g$; *and there are maps of every degree $m \in \mathbb{Z}$.*

(b) *If M is nonorientable and $\partial M = \varnothing$, then $f \simeq g$ if and only if* $\deg_2 f = \deg_2 g$; *and there are maps of every degree $m \in \mathbb{Z}_2$.*

(c) *If $\partial M \neq \varnothing$ then $f \simeq g$.*

Proof. We first show that there are maps of every degree. Let M be as in (a). The constant map $M \to S^n$ has degree 0. Given $m \in \mathbb{Z}_+$ let $\varphi_i: U_i \to \mathbb{R}^n$, $i = 1, \ldots, n$ be disjoint surjective charts which preserve orientation. Let $s: \mathbb{R}^n \to S^n - P$ be the inverse of stereographic projection from the north pole P so that s preserves orientation. Define

$$f: M \to S^n,$$

$$f = \begin{cases} s\varphi_i & \text{on} \quad U_i \\ \text{constant map } P & \text{on} \quad M - \cup U_i. \end{cases}$$

Then f is continuous and has degree m. If the φ_i were orientation reversing f would have degree $-m$. Taking $m = 1$ and ignoring orientations proves the second part of (b).

The first parts of (a) and (b) are consequences of Theorems 1.8 and 1.9 with $W = M \times I$. To prove (c) let M' be the double of M, two copies of M glued along ∂M, and $p: M' \to M$ the map identifying the two copies, and $i: M \to M'$ the embedding of one copy. It is easy to see that $fp: M' \to S^n$ has degree 0 if M is orientable, and otherwise $\deg_2(fp) = 0$. Therefore fp is homotopic to a constant map c (which also has degree 0). Since $fpi = f$, it follows that $f \simeq c$. Similarly $g \simeq c$, so $f \simeq g$.

QED

Exercises

1. A complex polynomial of degree n defines a map of the Riemann sphere to itself of degree n. What is the degree of the map defined by a rational function $p(z)/q(z)$?

2. (a) Let M, N, P be compact connected oriented n-manifolds without boundaries and $M \xrightarrow{g} N \xrightarrow{f} P$ continuous maps. Then $\deg(fg) = (\deg g)(\deg f)$. The same holds mod 2 if M, N, P are not oriented.

(b) The degree of a homeomorphism or homotopy equivalence is ± 1.

***3.** Let $\mathfrak{M}_n$ be the category whose objects are compact connected n-manifolds and whose morphisms are homotopy classes $[f]$ of maps $f: M \to N$. For an object M let $\pi^n(M)$ be the set of homotopy classes $M \to S^n$. Given $[f]: M \to N$ define $[f]^*: \pi^n(N) \to \pi^n(M)$, $[f]^*[g] = [gf]$. This makes π^n a contravariant functor from $\mathfrak{M}_n$ to the category of sets.

(a) There is a unique way of lifting this functor to the category of groups so that $\pi^n(S^n) = \mathbb{Z}$ with the identity map corresponding to $1 \in \mathbb{Z}$.

(b) Given the group structure of (a), for each M there is an isomorphism

$$\pi^n(M) \approx \begin{cases} \mathbb{Z} & \text{if} \quad M \text{ is orientable, } \partial M = \emptyset \\ \mathbb{Z}_2 & \text{if} \quad M \text{ is nonorientable, } \partial M = \emptyset \\ 0 & \text{if} \quad \partial M \neq \emptyset. \end{cases}$$

But there is no *natural* family of such isomorphisms.

4. A continuous map $f: S^n \to S^n$ such that $f(x) = f(-x)$ has even degree.

***5.** Let M, N be compact connected oriented n-manifolds, $\partial M = \emptyset$.

(a) Suppose $n \geqslant 2$. If there exists a map $S^n \to M$ of degree one, then M is simply connected. More generally:

(b) If $f: M \to N$ has degree 1 then the induced homomorphism of fundamental groups $f_{\#}: \pi_1(M) \to \pi_1(N)$ is surjective.

(c) If $f: M \to N$ has degree $k \neq 0$ then the image of $f_{\#}$ is a subgroup whose index divides $|k|$.

6. Let $M \subset \mathbb{R}^{n+1}$ be a compact n-dimensional submanifold, $\partial M = \emptyset$. For each $x \in \mathbb{R}^{n+1} - M$ define

$$\sigma_x: M \to S^n, \qquad y \mapsto (y - x)/|y - x|.$$

Then x and y are in the same component of $\mathbb{R}^{n+1} - M$ if and only if $\sigma_x \simeq \sigma_y$, and x is in the unbounded component if and only if $\sigma_x \simeq$ constant. If M is connected then x is in the bounded component if and only if $\deg(\sigma_x) = \pm 1$.

7. Let $M, N \subset \mathbb{R}^q$ be compact oriented submanifolds without boundaries, of dimensions m, n respectively. Assume that M and N are disjoint, and $m + n = q - 1$. The *linking number* $\mathrm{Lk}(M,N)$ is the degree of the map

$$M \times N \to S^{q-1}$$
$$(x, y) \mapsto (x - y)/|x - y|.$$

Then:

(a) $\mathrm{Lk}(M,N) = (-1)^{(m-1)(n-1)} \mathrm{Lk}(N,M)$.

(b) If M can be deformed to a point in $M - N$, or bounds an oriented compact submanifold in $M - N$, then $\mathrm{Lk}(M,N) = 0$.

(c) Let S, S' be the boundary circles of a cylinder embedded in $\mathbb{R}^3$ with k twists. Then, with suitable orientations, $\mathrm{Lk}(S,S') = k$.

(d) Let C_1 and $C_2 \subset \mathbb{R}^3$ be cylinders embedded with k_1 and k_2 twists respectively. If $|k_1| \neq |k_2|$ there is no diffeomorphism of $\mathbb{R}^3$ carrying C_1 onto C_2.

8. Let $M \subset \mathbb{R}^{n+1}$ be a compact n-dimensional submanifold without boundary. Two points $x, y \in \mathbb{R}^{n+1} - M$ are separated by M if and only if $\mathrm{Lk}(\{x,y\},M) \neq 0$. (See Exercise 7.)

9. The *Hopf invariant* of a map $f:S^3 \to S^2$ is defined to be the linking number $H(f) = \mathrm{Lk}(g^{-1}(a),g^{-1}(b))$ (see Exercise 7) where g is a C^∞ map homotopic to f and a, b are distinct regular values of g. The linking number is computed in

$$\mathbb{R}^3 = S^3 - c, \qquad c \neq a, b.$$

(a) $H(f)$ is a well-defined homotopy invariant of f which vanishes if f is null homotopic.

(b) If $g:S^3 \to S^3$ has degree p then $H(fg) = pH(f)$.

(c) If $h:S^2 \to S^2$ has degree q then $H(hf) = q^2H(f)$.

(d) Let $S^3 \subset \mathbb{C}^2$ be the unit sphere and $S^2 = CP^1$. The *Hopf map*

$$\varphi:S^3 \to S^2,$$
$$\varphi(z,w) = [z,w]$$

has Hopf invariant 1. Hence φ is not null homotopic.

10. Let U, V be noncompact oriented n-manifolds without boundaries and $h:U \to V$ a *proper* C^∞ map. The degree of h is defined as usual,

$$\deg h = \textstyle\sum_x \deg_x h \qquad (x \in h^{-1}(y))$$

where y is a regular value.

(a) $\deg h$ is independent of y, and if g is a C^∞ map homotopic to h by a *proper* homotopy $U \times I \to V$ then $\deg g = \deg h$. Thus the degree of any continuous proper map $f:U \to V$ can be defined by choosing h sufficiently close to f.

(b) In particular the degree of a homeomorphism $U \to V$ is defined; it is always ± 1. (Compare Exercise 2).

(c) A topological n-manifold without boundary is called *topologically orientable* if it has an atlas whose coordinate changes have degree $+1$ on each component. A smooth manifold is orientable if and only if it is topologically orientable.

(d) Orientability of a smooth manifold is a topological invariant.

11. The fundamental theorem of algebra can be generalized as follows. Let $U \subset \mathbb{R}^n$ be a nonempty open set and $f:U \to \mathbb{R}^n$ a C^1 map. Assume: (a) f is proper; (b) outside some compact set, $\mathrm{Det}(Df_x)$ does not change sign and is not identically zero. Then f is surjective. In particular the equation $f(x) = 0$ has a solution.

12. Let $f_1, \ldots, f_n$ be real [or complex] polynomials in $n \geqslant 2$ variables. Write $f_k = h_k + r_k$ where h_k is a homogeneous polynomial of degree $d_k > 0$ and r_k has smaller degree. Assume that $x = (0, \ldots, 0)$ is the only solution to $h_1(x) = \cdots = h_n(x) = 0$. Assume also that $\mathrm{Det}[\partial h_i/\partial x_j] \neq 0$ at all nonzero x in $\mathbb{R}^n$ [or $\mathbb{C}^n$]. Then the system of equations $f_k(x) = 0$, $k = 1, \ldots, n$, has a solution in $\mathbb{R}^n$ [or $\mathbb{C}^n$]. [Hint: Exercise 11.]

2. Intersection Numbers and the Euler Characteristic

Let W be an oriented manifold of dimension $m + n$ and $N \subset W$ a closed oriented submanifold of dimension n. Let M be a compact oriented m-manifold. Suppose $\partial M = \partial N = \varnothing$.

Let $f:M \to W$ be a C^∞ map transverse to N.

A point $x \in f^{-1}(N)$ has *positive* or *negative type* according as the composite linear isomorphism

$$M_x \xrightarrow{Tf} W_y \to W_y/N_y, \qquad y = f(x)$$

preserves or reverses orientation; we write $\#_x(f,N) = 1$ or -1, respectively. The *intersection number* of (f,N) is the integer

$$\#(f,N) = \sum \#_x(f,N),$$

summed over all $x \in f^{-1}(N)$.

2.1. Theorem. *If $f, g: M \to W$ are homotopic C^∞ maps transverse to N then $\#(f,N) = \#(g,N)$.*

Proof. The proof is similar to that of Theorem 1.2 and is left to the reader.

QED

For any continuous map $g: M \to W$ we define $\#(g,N) = \#(f,N)$ where f is a C^∞ map which is transverse to N and homotopic to g. By Theorem 2.1, $\#(g,N)$ is well defined.

Note that it is not really necessary for N and W to be oriented; all that is actually used is an orientation of the normal bundle of N.

If M is also a submanifold of W and $i: M \to N$ is the inclusion, the *intersection number* of (M,N) is the integer $\#(M,N) = \#(i,N)$. We put $\#(M,N) = \#(M,N; W)$ to emphasize W. If both M and N are compact then $\#(M,N) = (-1)^{mn}\#(N,M)$ as is easily proved.

Clearly $\#(f,N) = 0$ if f is homotopic in W to $g: M \to W - N$. In particular *if M and N are closed submanifolds of $\mathbb{R}^{m+n}$ with M compact and $\partial M = \partial N = \varnothing$, then $\#(M,N; \mathbb{R}^{m+n}) = 0$.*

It is not generally true that if $\#(M,N) = 0$ then M is deformable into $W - N$; Figure 5–3 shows a counterexample of two circles on a surface S of genus 2. If M, N, and W are allowed to have boundaries, $\#(f,N)$ and $\#(M,N)$ are defined in the same way whenever M and N are neat submanifolds and $\partial M \cap \partial N = \varnothing$. In this case the C^∞ map $g: M \to N$

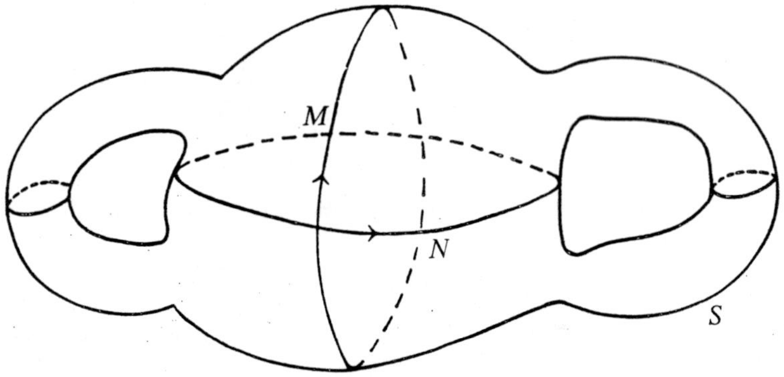

Figure 5–3. $\#(M,N) = 0$, but M is not deformable into $S - N$.

homotopic to f must be chosen to an approximate f so closely that there is a homotopy from f to g in $W - \partial N$. In this case $\#(f,N)$ is only an invariant of this kind of homotopy.

If M, N or W is not oriented, mod 2 intersection numbers $\#_2(f,N)$ and $\#_2(M,N)$ are defined in a similar way.

Let $\xi = (p,E,M)$ be an n-dimensional oriented vector bundle over M; as usual M is identified with the zero section. Assume M is connected, compact, n-dimensional and without boundary. The *Euler number* of ξ is defined to be the integer

$$X(\xi) = \#(M,M) = \#(M,M; E)$$

To compute $X(\xi)$ approximate the zero section $Z:M \to E$ by a C^∞ map h transverse to $Z(M)$. If the approximation is close enough then $ph:M \to N$ is a diffeomorphism and we obtain a C^∞ *section* g transverse to the zero section by setting $g = h(ph)^{-1}$: thus $fg = 1_M$.

Let $x_1, \ldots, x_m \in M$ be the zeros of g. Let $\varphi_i:\xi|U_i \to U_i \times \mathbb{R}^n$ be local trivializations of ξ over open sets $U_i \subset M$ such that $x_i \in U_i$. The composition

$$F_i: U_i \xrightarrow{g} E|U_i \to U_i \times \mathbb{R}^n \to \mathbb{R}^n$$

has $0 \in \mathbb{R}^n$ as a regular value and $x_i \in F_i^{-1}(0)$. Then

$$\deg_{x_i} F_i = \#_{x_i}(g,M).$$

Hence

$$X(\xi) = \textstyle\sum_i \deg_{x_i} F_i.$$

We define the *index at x_i of the section g* to be $\#_{x_i}(g,M)$.

If $\xi = TM$ then $X(\xi)$ is called the *Euler characteristic* of M, denoted by $\chi(M)$. Later we shall define $\chi(M)$ for nonorientable and ∂-manifolds.

2.2. Theorem. *If ξ (as above) has a section f that is nowhere zero then* $X(\xi) = 0$.

Proof. f is homotopic to the zero section by the linear homotopy $(x,t) \to tf(x)$. Hence if $i:M \to E$ is the zero section,

$$X(\xi) = \#(i,M) = \#(f,M) = 0.$$

QED

To compute $\chi(M)$ one can start with a C^∞ vector field[1] $f:M \to TM$ transverse to the zero section. At each zero x_i of f let $\varphi_i: U_i \to \mathbb{R}^n$ be a chart (preserving orientation). Then $T\varphi_i \circ f \circ \varphi_i^{-1}$ is a C^∞ vector field on $\varphi_i(U_i)$ and thus defines a C^∞ map $g_i:\varphi_i(U_i) \to \mathbb{R}^n$ with a regular value at 0. Denote by d_i the degree of g_i at $f(x_i)$; then $\chi(M) = \sum d_i$. The integer d_i is the *index at x_i of the vector field f*; it is independent the choice of (φ_i, U_i) and the orientation of M. We denote d_i by $\operatorname{Ind}_{x_i} f$.

[1] A *vector field* is a section of the tangent bundle.

As an example we compute $\chi(S^n)$. Let P be the north pole and $Q = -P$ the south pole. Let

$$\sigma: S^n - P \to \mathbb{R}^n$$
$$\tau: S^n - Q \to \mathbb{R}^n$$

be the stereographic projections. The coordinate change

$$\tau\sigma^{-1} = \sigma\tau^{-1}: \mathbb{R}^n - 0 \to \mathbb{R}^n - 0$$

is given by $x \mapsto x/|x|^2$.

Let f be the vector field on $S^n - P$ whose representation via σ is the identity vector field on $\mathbb{R}^n$. Then $f(x) \to 0$ as $x \to P$ and we define $f(P) = 0$. Thus $f: S^n \to TS^n$ has zeroes only at Q and P.

In τ coordinates f corresponds to the vector field $x \mapsto -x$ on $S^n - (Q)$. Thus f is C^∞.

The identity map of $\mathbb{R}^n$ has degree 1 at 0, the antipodal map has degree $(-1)^n$. Therefore

$$\operatorname{Ind}_P f = 1, \qquad \operatorname{Ind}_Q f = (-1)^n.$$

Thus we have proved:

2.3. Theorem.

$$\chi(S^n) = 1 + (-1)^n = \begin{cases} 2 & \text{if} \quad n \text{ is even,} \\ 0 & \text{if} \quad n \text{ is odd.} \end{cases}$$

2.4. Corollary. *Every vector field on S^{2n} vanishes somewhere.*

Some other computations are given by:

2.5. Theorem. (a) *Let M and N be compact oriented manifolds without boundaries. Then $\chi(M \times N) = \chi(M)\chi(N)$.*

(b) *Let ξ be an n-dimensional oriented vector bundle over a compact oriented n-manifold M^n without boundary. Then $X(\xi) = 0$ if n is odd.*

(c) *$\chi(M) = 0$ if M is an odd dimensional compact oriented manifold without boundary.*

Proof. (a) is proved by choosing vector fields f, g on M, N and using $f \times g: M \times N \to TM \times TN$ to compute $\chi(M \times N)$. The details are left as an exercise.

(b) is proved by using sections f and $-f$ to compute $X(\xi)$ in two ways. One finds that

$$\operatorname{Ind}_x f = (-1)^n \operatorname{Ind}_x(-f), \qquad n = \dim M.$$

Hence if n is odd, $X(\xi) = -X(\xi)$. And (c) follows from (b).

QED

Now we define $\chi(M)$ for a nonorientable compact manifold M, $\partial M = \varnothing$. Let $f: M \to TM$ be a section transverse to M. For each $x \in M$ there is a canonical identification

$$(TM)_x/M_x \approx M_x;$$

see Theorem 4.2.1. Therefore Tf induces an automorphism Φ_x via the composition

$$\Phi_x : M_x \xrightarrow{Tf} (TM)_x \to TM_x/M_x \approx M_x.$$

Define

$$\operatorname{Ind}_x f = \operatorname{Det} \Phi_x / |\operatorname{Det} \Phi_x|,$$
$$\chi_f(M) = \sum \operatorname{Ind}_x f, \qquad x \in f^{-1}(M).$$

This definition of course makes sense if M is oriented, and in that case it is easy to check that it is the same as $\chi(M)$. In particular, it is independent of f and of the orientation in this case.

For M nonorientable, let $p: \tilde{M} \to M$ be the oriented double covering. Given f as above let $\tilde{f}: \tilde{M} \to T\tilde{M}$ be the unique section covering f. Let $\tilde{x} \in \tilde{f}^{-1}(\tilde{M})$; put $p(\tilde{x}) = x$. It is clear that

$$\operatorname{Ind}_{\tilde{x}} \tilde{f} = \operatorname{Ind}_x f.$$

It follows that

$$\chi_{\tilde{f}}(\tilde{M}) = 2x_f(M),$$

or

$$\tfrac{1}{2}\chi(\tilde{M}) = \chi_f(M).$$

Thus $\chi_f(M)$ is independent of f.

Parts (a) and (c) of Theorem 2.5 hold for M nonorientable.

The Euler characteristic $\chi(M)$ of a compact ∂-manifold M is defined as follows. Let $f: M \to TM$ be a section which is transverse to the zero section M, and such that f points *outward* at points of ∂M. Such a section always exists; for example, an outward section over ∂M can be obtained from a collar, extended over M by a partition of unity, and then be made transverse to the zero section by approximation. Moreover any two such *outward vector fields* are connected by a homotopy of outward vector fields. We define

$$\chi(M) = \chi_f(M) = \sum_x \operatorname{Ind}_x f, \qquad x \in f^{-1}(M).$$

To show that $\chi_f(M)$ is independent of f, let $\tilde{M} \to M$ be the oriented double covering of M and let $\tilde{f}$ be the vector field on $\tilde{M}$ which covers f. Then $\tilde{f}$ is outward, and local computations show that $\chi_{\tilde{f}}(\tilde{M}) = 2\chi_f(M)$. It therefore suffices to show that $\chi_f(M)$ is independent of f when M is oriented; and this proof is similar to that of Theorem 1.2.

The Euler characteristic $\chi(M)$ is thus defined for any compact manifold M. Note that Theorem 2.5 (a) is true whenever ∂M or ∂N is empty.

The proof of the following lemma is left to the reader (a similar argument was given in the proof of Theorem 1.4).

2.6. Lemma. *Let M be a connected manifold, $U \subset M$ an open set $F \subset M$ a finite set. Then there is a diffeomorphism of M carrying F into U.*

Using the lemma we prove:

2.7. Theorem. *Let M be a compact connected ∂-manifold. Then M has a nonvanishing vector field.*

Proof. Let M' be the double of M, containing M as a submanifold. Then M' has a vector field f with a finite set F of zeros. Let $\varphi: M' \to M'$ be a diffeomorphism taking F into $M' - M$. Then $T\varphi \circ f \circ \varphi^{-1}|M$ is a vector field on M without zeros!

QED

The following lemma relates the index of a zero of a vector field to the degree of a map into a sphere.

2.8. Lemma. *Let $D \subset \mathbb{R}^n$ be an n-disk with center x. Let $U \subset \mathbb{R}^n$ be an open set containing D and $f: U \to \mathbb{R}^n$ a C^∞ map, considered as a vector field on U. Suppose 0 is a regular value and $x \in D \cap f^{-1}(0)$. Define a map*

$$g: \partial D \to S^{n-1},$$
$$y \to f(y)/|f(y)|.$$

Then $\deg g = \operatorname{Ind}_x f$.

Proof. We may suppose for simplicity that $x = 0$. Define $f_t: U \to \mathbb{R}^n$ by:

$$f_t(y) = \begin{cases} t^{-1}y(ty), & 1 \geq t > 0 \\ Df_0(y), & t = 0; \end{cases}$$

define $g_t: \partial D \to S^{n-1}$ like g, using f_t instead of f. Since g_0 is homotopic to $g_1 = g$, it follows that $\deg g_0 = \deg g$.

We claim $\operatorname{Ind}_0 f_0 = \operatorname{Ind}_0 f$. This is because the map

$$\Phi: D \times I \to \mathbb{R}^n \times I,$$
$$(y,t) \to (f_t(y),t)$$

is C^∞ (see proof of Theorem 4.4.3) and $T\Phi$ thus induces a homotopy between $Df_0(0)$ and $Df_1(0)$.

It remains to prove $\operatorname{Ind}_0 f_0 = \deg g_0$. Now f_0 is linear, and hence is homotopic through linear maps to an element of $O(n)$. Thus it suffices to prove the lemma for the special case where f is orthogonal. But then $g = f$ and in this case the lemma is easy to verify: $\operatorname{Ind}_0 f = \pm 1 = \deg g$ according as f preserves or reverses orientation.

QED

The degree of $g: \partial D \to S^{n-1}$ in Theorem 2.8 is defined whenever x is an *isolated* zero of $f: U \to \mathbb{R}^n$, even if f is merely continuous. Moreover $\deg g$ is independent of D, by Theorem 1.6(b). This permits us to extend significantly the definition of $\operatorname{Ind}_x f$. Let $U \subset \mathbb{R}^n$ be an open set, $f: U \to \mathbb{R}^n$ a vector field on U and $x \in U$ an isolated zero of f. The *index of f at x* is

$\mathrm{Ind}_x f = \deg g$, where for some n-disk $D \subset U$ with center x,

$$g : \partial D \to S^{n-1},$$
$$g(y) = f(y)/|f(y)|.$$

If 0 is a regular value of f this agrees with the older definition. *Note that* $\deg g$ *can take any integer value.*

2.9. Lemma. *Let $W \subset \mathbb{R}^n$ be a connected compact n-dimensional submanifold and $f: W \to \mathbb{R}^n$ a C^∞ map. Assume $f^{-1}(0)$ is a finite subset of $W - \partial W$. If*

$$\sum \mathrm{Ind}_x f = 0, \qquad (x \in f^{-1}(0)),$$

there is a map $g: W \to \mathbb{R}^n - 0$ which equals f on ∂W; and conversely.

Proof. Let $f^{-1}(0) = \{x_1, \ldots, x_k\}$. Let $D_1, \ldots, D_k$ be small disjoint n-disks in Int W, centered at $x_1, \ldots, x_k$ respectively.

Let $W_0 = W - \cup \operatorname{Int} D_i$. Then $\partial W_0 = \partial W \bigcup_{i=1}^{n} \partial D_i f(W_0) \subset \mathbb{R}^n - 0$.

Notice that $\deg(f|\partial W_0) = 0$ by 1.8. Define

$$g: W_0 \to S^{n-1},$$
$$y \mapsto f(y)/|f(y)|.$$

According to the preceding lemma,

$$\sum_{i=1}^{k} \deg(g|\partial D_i) = 0.$$

It follows that $\deg(g|\partial W) = 0$; by Theorem 1.8 there is a map $W \to S^{n-1}$ extending $g|\partial W$. The composition

$$\partial W \xrightarrow{G} S^{n-1} \to \mathbb{R}^n - 0$$

is homotopic to $f|\partial W$. Since g extends over W, so does $f|\partial W$. The converse part of Theorem 2.9 is left to the reader.

QED

We can now prove the converse of Theorem 2.2.

2.10. Theorem. *Let M be a compact connected oriented n-manifold without boundary and ξ an oriented n-plane bundle over M. If $X(\xi) = 0$ then ξ has a nonvanishing section.*

Proof. Let $f: M \to TM$ be a C^∞ section tranverse to M. By Theorem 2.6 we may assume that the finite set $f^{-1}(M)$ lies in the domain of the chart $\varphi: V \approx \mathbb{R}^n$. The map

$$T\varphi \circ f \circ \varphi^{-1}: U \to TU = U \times \mathbb{R}^n$$

is a vector field on $\mathbb{R}^n$ transverse to the zero section, or what is the same thing, a map $g:\mathbb{R}^n \to \mathbb{R}^n$ transverse to 0. The assumption $X(\xi) = 0$ implies that

$$\sum_x \operatorname{Ind}_x g = 0 \qquad (x \in g^{-1}(0)).$$

Let $B \subset \mathbb{R}^n$ be an n-disk containing $g^{-1}(0)$ in its interior. By Theorem 2.9 there is a map $h:\mathbb{R}^n \to \mathbb{R}^n - 0$ which equals g on $\mathbb{R}^n - \operatorname{Int} B$. Thus h defines a nonvanishing vector field g_1 on $\mathbb{R}^n$ which equals g on $\mathbb{R}^n - \operatorname{Int} B$. Define

$$f_1 : M \to TM,$$

$$f_1 = \begin{cases} f & \text{on} \quad M - \varphi^{-1}(\operatorname{Int} B) \\ T_U\varphi^{-1} \circ g_1 \circ \varphi & \text{on} \quad B. \end{cases}$$

Then f_1 is nonvanishing.

QED

We use all the preceding results to prove a classical theorem of Whitney [3]. We follow Whitney's proof.

2.11. Theorem. *Let $M \subset \mathbb{R}^{2n}$ be a compact oriented n-dimensional submanifold without boundary. Then M has a nonvanishing normal vector field.*

Proof. We may assume M connected. Let ν be the normal bundle of M. By Theorem 2.10 it suffices to prove that $X(\nu) = 0$. We identify a neighborhood W of the zero section of ν with a neighborhood of M in $\mathbb{R}^n$. Then

$$X(\xi) = \#(M,M; W) = \#(M,M; \mathbb{R}^n) = 0.$$

QED

If M in Theorem 2.11 is not assumed orientable, the conclusion may be false, as Whitney showed for $P^2 \subset \mathbb{R}^4$.

Exercises

1. Consider complex projective m-space CP^m as a submanifold of CP^n in the natural way. Then (with natural orientations)

$$\#(CP^m, CP^{n-m}) = 1.$$

A similar result holds mod 2 for real projective spaces.

2. $\#(f,N; W) = 0$ if $f:M \to W$ extends to an oriented compact manifold bounded by M, or if N bounds a closed oriented submanifold of W, or if f or N is null homotopic in W. Similarly for $\#_2(f,N)$.

3. Let M be a compact oriented manifold without boundary. Let

$$M_\Delta = \{(x,x) \in M \times M\}$$

be the diagonal submanifold.

(a) $\chi(M) = \#(M_\Delta, M_\Delta)$

(b) $\chi(M) = 0$ if and only if there is a map $f:M \to M$ without fixed point, which is homotopic to the identity. This is true even if M is nonorientable.

4. Let M, N, W be oriented manifolds without boundaries, $\dim M + \dim N = \dim W$. The *intersection number of maps* $f:M \to W$, $g:N \to W$ is defined as

$$\#(f,g) = \#(f \times g, W_\Delta; W \times W).$$

This integer depends only on the homotopy classes of f and g. If g is an embedding then

$$\#(f,g) = \#(f,g(N)).$$

In general

$$\#(g,f) = (-1)^{\dim M \dim N}\, \#(f,g).$$

For unoriented manifolds a mod 2 intersection number is defined similarly.

5. Let $\xi_i = (p_i,E_i,M_i)$, $i = 0, 1$, be oriented n-plane bundles over compact oriented n-manifolds without boundary. Put

$$\xi_0 \times \xi_1 = (p_0 \times p_1, E_0 \times E_1, M_0 \times M_1).$$

Then

$$X(\xi_0 \times \xi_1) = X(\xi_0)X(\xi_1).$$

6. Let ξ be an n-plane bundle over a connected k-manifold M.

(a) If $k < n$, ξ has a nonvanishing section.

(b) If $k = n$ and $x \in M$, ξ has a section which vanishes only at x.

(c) If $k = n$, and $\partial M \neq \varnothing$ or M is not compact, then ξ has a nonvanishing section.

7. (a) Suppose a compact n-manifold can be expressed as $A \cup B$ where A, B are compact n-dimensional submanifolds and $A \cap B$ is an $(n-1)$ − dimensional submanifold. Then $\chi(A \cap B) = \chi(A) + \chi(B) - \chi(A \cap B)$.

(b) $\chi(\partial A)$ is even. [Hint: take $B \approx A$.]

8. The Euler characteristic of an (orientable) surface of genus g is $2 - 2g$. [Use Exercise 7.]

9. (a) Let M be a possibly nonorientable compact manifold without boundary, and $G:M \to M \times M$ a C^∞ map. Then the integer $L(G) = \#(G,M_\Delta; M \times M)$ is well-defined using arbitrary local orientations of M at points x where $G(x) \in M_\Delta$, and the corresponding local orientations of $M \times M$ and M_Δ at (x,x). Moreover $L(G)$ is a homotopy invariant and so is defined for continuous maps G. If $g:M \to M$ is any continuous map, the *Lefschetz number* of g is the integer $\mathrm{Lef}(g) = L(G)$ where $G(x) = (x,g(x))$. $\mathrm{Lef}(g)$ is a homotopy invariant of g. The Lefschetz number of 1_M is $\chi(M)$. If $\mathrm{Lef}(g) \neq 0$ then g has a fixed point.

10. Let $x \in M$ be an isolated fixed point of a continuous map $g:M \to M$. Let $\varphi: U \to \mathbb{R}^n$ be a chart at x; put $\varphi(x) = y$. The vector field $f(z) = \varphi g \varphi^{-1}(z) - z$ is defined on a neighborhood of y in $\mathbb{R}^n$ and has y for an isolated zero. Define

$$\mathrm{Lef}_x(g) = \mathrm{Ind}_y f.$$

This is independent of (φ,U). If φ is C^1 then

$$\mathrm{Lef}_x(g) = \mathrm{Det}(T_x\varphi - I), \qquad I = \text{identity map of } M_x.$$

If the set Fix (g) of fixed points of g is finite, then (see Exercise 9)

$$\mathrm{Lef}(g) = \sum_x \mathrm{Lef}_x(g) \qquad (x \in \mathrm{Fix}(g)).$$

11. Every continuous map $P^{2n} \to P^{2n}$ has a fixed point. [Consider maps of S^{2n} which commute with the antipodal map. See Exercises 9, 10.]

12. Lemma 2.9 is true even if the map f is merely continuous.

13. (a) There is a continuous map $f:S^2 \to S^2$ of degree 2, that has exactly two periodic points. (A point x is *periodic* if $f^n(x) = x$ for some $n > 0$.)

**(b) If $f:S^2 \to S^2$ is C^1 and has degree d such that $|d| > 1$, then f has infinitely many periodic points. (Shub–Sullivan [1]).

***(c) Identify the 2-torus T^2 with the coset space $\mathbb{R}^2/\mathbb{Z}^2$ and let $f:T^2 \to T^2$ be the diffeomorphism induced from the linear operator on $\mathbb{R}^2$ whose matrix is $\left[\begin{smallmatrix} 2 & 1 \\ 1 & 1 \end{smallmatrix}\right]$. Shub and Sullivan ask: if $g:T^2 \to T^2$ is a continuous map homotopic to f, must g have infinitely many periodic points?

14. Let M, N be compact oriented n-manifolds without boundary; assume N is connected. Then the degree of a map $f:M \to N$ equals the intersection number of the graph of f with $M \times y$ in $M \times N$, for any $y \in N$.

***15.** Using intersection numbers and some elementary homotopy theory one can prove that every diffeomorphism of the complex projective plane CP^2 preserves orientation. Using the fact that $\pi_2(CP^2)$ is infinite cyclic, generated by the natural inclusion $i:S^2 = CP^1 \subset CP^2$, one sees that if $h:CP^2 \approx CP^2$ then $h_\#[i] = \pm[i]$ in $\pi_2(CP^2)$. Therefore

$$\#(hi,hi) = \#(i,i) = 1$$

(see Exercise 4). On the other hand it is easy to see that for any maps

$$f,g:S^2 \to CP^2,$$
$$\#(hf,hg) = (\deg h)[\#(f,g)].$$

Therefore $\deg h = 1$.

16. Theorem 2.11 generalizes as follows. Let $M^n \subset N^{2n}$ be a compact submanifold. Suppose M^n and N^{2n} are orientable and $M^n \simeq 0$ in N^{2n}. Then M^n has a nonvanishing normal vector field (for any Riemannian metric on N^{2n}).

17. What is the Euler number of the normal bundle of CP^n in CP^{2n}?

$$[z_0, \ldots, z_n] \mapsto [w_0, \ldots, w_n]$$

where $w_j = (\sum_k A_{jk}z_k)^p$, if p is an integer and $[A_{jk}] \in GL(n,\mathbb{C})$?

18. What is the Euler number of the normal bundle of CP^n in CP^{2n}?

19. Let $\xi \to S^n$ be an orthogonal oriented n-plane bundle.

(a) ξ corresponds to an element $\alpha \in \pi_{n-1}(SO(n))$. (Compare Exercise 8, Section 4.3.)

(b) $X(\xi)$ is the image of α in $\pi_{n-1}(S^{n-1}) = \mathbb{Z}$ by the homomorphism

$$f_\#:\pi_{n-1}(SO(n)) \to \pi_{n-1}(S^{n-1})$$

induced by the map $f:SO(n) \to S^{n-1}$, where f is defined by evaluation on the north pole $P \in S^{n-1}$.

20. Verify the statement of Heegard quoted at the beginning of Chapter 4.

3. Historical Remarks

The origin of the notion of the degree is Kronecker's "characteristic of of a system of functions" defined in 1869. The type of problem Kronecker studied was the following (in modern terminology). Let $F_0, \ldots, F_n$ be C^1 maps from $\mathbb{R}^n$ to $\mathbb{R}$. Suppose $0 \in \mathbb{R}$ is a regular value for each of them, and that each submanifold $M_i = F_i^{-1}(-\infty,0]$ is compact. If there is no common zero of $F_0, \ldots, F_n$ on ∂M_i for $i = 1, \ldots, n$, how many common zeros are there for the n maps $F_j: M_i \to \mathbb{R}, j = 0, \ldots, i-1, i+1, \ldots, n$? Kronecker gave an integral formula which is equivalent to the degree of the composite map

$$\partial M_i \xrightarrow{G_i} \mathbb{R}^n - 0 \xrightarrow{\pi} S^{n-1}$$

where $G_i = (F_0, \ldots, \hat{F}_i, \ldots, F_n)$ and $\pi(x) = x/|x|$. He proved that this degree is independent of i and equals the algebraic number of zeros of G_i in M_i. He also showed that the total curvature of a compact surface $M^2 \subset \mathbb{R}^3$ is 2π times the degree of the Gauss map $M^2 \to S^2$. Later Walter van Dyck showed that the degree of the Gauss map equals the Euler characteristic of M, thus giving the first proof of what is now wrongly called the Gauss–Bonnet theorem.

An influential and still interesting article by Hadamard [1] in 1910 gave a more geometrical presentation of Kronecker's ideas. Kronecker's work is discussed in modern terms in the books by Lefschetz [1], and Alexandroff and Hopf [1].

The topological idea of the degree of a map is due to Brouwer [1]. Brouwer made fundamental contributions to the topology of manifolds (see Lefschetz [1] for an extensive bibliography). In his later years, however, he developed the intuitionistic view of mathematics and repudiated some of his earlier results.

Our treatment of the degree closely follows that of Pontryagin [1].

Chapter 6

Morse Theory

La topologie est précisément la discipline mathématique qui permet la passage du local au global.

—R. Thom, *Stabilité Structurelle et Morphogenèse*, 1972

Up to this point we have obtained results of a very general nature: all n-manifolds embed in $\mathbb{R}^{2n+1}$, all maps can be approximated by C^∞ maps, etc. These are useful tools but they give no hint as to how to analyze a particular manifold, or class of manifolds. As yet we are unable even to classify compact 2-manifolds.

In this chapter we analyze the level sets $f^{-1}(y)$ of a function $f: M \to \mathbb{R}$ having only the simplest possible critical points. Such a function is called a "Morse function." The decomposition of M into these level sets contains an amazing amount of information about the topology of M. For example we will show in Section 6.4 how a CW-complex, homotopy equivalent to M, can be obtained from any Morse function. In Section 6.3 the Morse inequalities are proved. These relate the critical points of f to the homology groups of M; in particular they compute the Euler characteristic of M from any Morse function on M.

Morse functions are shown in Section 6.1 to be open and dense in $C_S^r(M,\mathbb{R})$, $2 \leqslant r \leqslant \infty$. At each critical point a special kind of chart is constructed, making a Morse function look like a nondegenerate quadratic form. The index of this form is called the index of the critical point. These charts give a complete local analysis of the function.

In Section 6.2, which starts out with some facts about differential equations, the sets $f^{-1}[a,b]$ which contain no critical point are investigated. Under mild restrictions it is shown that $f^{-1}[a,b] \approx f^{-1}(a) \times [a,b]$.

Section 6.3 contains the heart of Morse theory. Suppose $f^{-1}[a,b]$ contains exactly one critical point, of index k. It turns out that up to homotopy equivalence, $f^{-1}[a,b]$ is obtained from $f^{-1}(a)$ by attaching a k-cell. This leads directly to the Morse inequalities, and to the construction of a CW-complex homotopy equivalent to M which has one k-cell for each critical point of index k.

We have presented only the very beginning of Morse theory. For the subject's important applications to such fields as differential geometry and the calculus of variations the reader should consult M. Morse [1], Milnor [3], Palais [1], or Smale [3].

1. Morse Functions

Let M be a manifold of dimension n. The *cotangent bundle* T^*M is defined like the tangent bundle TM using the dual vector space $(\mathbb{R}^n)^* = L(\mathbb{R}^n,\mathbb{R})$ instead of $\mathbb{R}^n$. More precisely, as a set $T^*M = \bigcup_{x\in M}(M_x^*)$ where $M_x^* = L(M_x,\mathbb{R})$. If (φ,U) is a chart on M, a *natural chart* on T^*M is the map

$$T^*U \to \varphi(U) \times (\mathbb{R}^n)^*$$

which sends $\lambda \in M_x^*$ to $(\varphi(x),\lambda\varphi_x^{-1})$. The projection map $p: T^* \to M$ sends M_x^* to x.

Let $f: M \to \mathbb{R}$ be a C^{r+1} map, $1 \leqslant r \leqslant \omega$. For each $x \in M$ the linear map $T_xf: M_x \to \mathbb{R}$ belongs to M_x^*. We write

$$T_xf = Df_x \in M_x^*.$$

Then the map

$$Df: M \to T^*M,$$
$$x \mapsto Df_x = Df(x)$$

is a C^r section of T^*M. The local representation of Df, in terms of a chart on M and the corresponding natural chart on T^*M, is a map from an open set in $\mathbb{R}^n$ to $(\mathbb{R}^n)^*$, of the form $x \mapsto Dg(x)$ where g is the local representation of f. Thus Df generalizes the usual differential of functions on $\mathbb{R}^n$.

A critical point x of f is a zero of Df, that is, $Df(x)$ is the zero of the vector space M_x^*. Thus the set of critical points of f is the counter-image of the submanifold $Z^* \subset T^*M$ of zeroes. Note that $Z^* \approx M$ and the codimension of Z^* is $n = \dim M$.

A critical point x of f is *nondegenerate* if Df is transverse to Z^* at x. If all critical points of f are nondegenerate f is called a *Morse function.* In this case the set of critical points is a closed discrete subset of M.

The idea behind the definition of nondegenerate critical point is this. By means of local coordinates, assume $M = \mathbb{R}^n$ and $x \in \mathbb{R}^n$ is a critical point for $f: \mathbb{R}^n \to \mathbb{R}$. It is easy to see that x is nondegenerate precisely when x is a regular point for $Df: \mathbb{R}^n \to (\mathbb{R}^n)^*$. Therefore as y varies in a small neighborhood of x, Df_y takes on every value in a neighborhood of 0 in $(\mathbb{R}^n)^*$ exactly once. Moreover as y moves away from x with nonzero velocity, Df_y moves away from 0 with nonzero velocity.

Let $U \subset \mathbb{R}^n$ be open and let $g: U \to \mathbb{R}$ be a C^2 map. It is easy to see that a critical point $p \in U$ is nondegenerate if and only if the linear map

$$D(Dg)(p): \mathbb{R}^n \to (\mathbb{R}^n)^*$$

is an isomorphism. Identifying $L(\mathbb{R}^n,(\mathbb{R}^n)^*)$ with $L^2(\mathbb{R}^n)$ in the natural way, we see that this is equivalent to the condition that the symmetric bilinear map $D^2g(p): \mathbb{R}^n \times \mathbb{R}^n \to \mathbb{R}$ be nondegenerate. In terms of coordinates this means that the $n \times n$ *Hessian matrix*

$$\left[\frac{\partial^2 g}{\partial x_i\, \partial x_j}(p)\right]$$

has rank n. This provides a criterion in local coordinates for a critical point of a map $M \to \mathbb{R}$ to be nondegenerate.

Let $p \in U$ be a critical point of $g: U \to \mathbb{R}$. The *Hessian* of g at the critical point p is the quadratic form $H_p f$ associated to the bilinear form $D^2 g(p)$; thus

$$H_p f(y) = D^2 g(p)(y,y)$$
$$= \sum_{i,j} \frac{\partial^2 g}{\partial x_i \, \partial x_j}(p) y_i y_j.$$

This form is invariant under diffeomorphisms in the following sense. Let $V \subset \mathbb{R}^n$ be open and suppose $h: V \to U$ is a C^2 diffeomorphism. Let $q = h^{-1}(p)$ so that q is a critical point of $gh: V \to \mathbb{R}$. Then the diagram commutes,

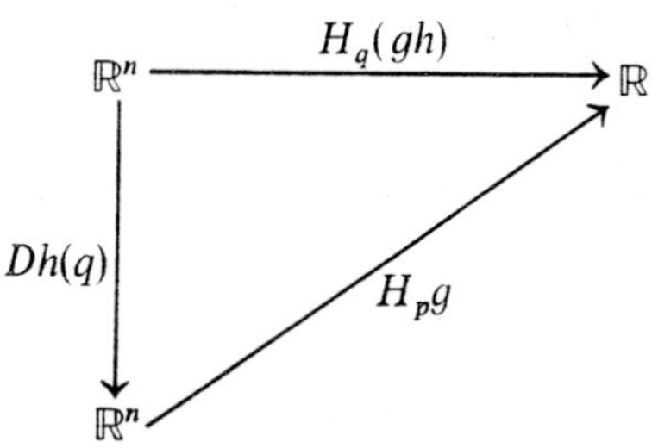

as a computation shows.

Now let $f: M \to \mathbb{R}$ be C^2. For each critical point x of f we define the Hessian quadratic form $H_x f: M_x \to \mathbb{R}$ to be the composition

$$H_x f: M_x \xrightarrow{D\varphi_x} \mathbb{R}^n \xrightarrow{H_{\varphi x}(f\varphi^{-1})} \mathbb{R}$$

where φ is any chart at x. The invariance property of Hessians of functions on $\mathbb{R}^n$ implies that $H_x f$ is well-defined independently of φ. Note that x is a nondegenerate critical point if and only if $H_x f$ is a nondegenerate quadratic form. Thus we obtain an alternate definition: a critical point of a C^2 real valued function is nondegenerate if and only if the associated Hessian quadratic form is nondegenerate.

Now let Q be a nondegenerate quadratic form on a vector space E. We say Q is *negative definite* on a subspace $F \subset E$ if $Q(x) < 0$ whenever $x \in F$ is nonzero. The largest possible dimension of a subspace on which Q is negative definite is the *index* of Q, denoted by Ind Q. If $A = [a_{ij}]$ is a symmetric $n \times n$ matrix expressing $Q(x)$ as $\sum a_{ij} x_i x_j$ for some choice of linear coordinates on E, then the index of Q equals the number of negative eigenvalues of A, counting multiplicities.

Let $p \in M$ be a nondegenerate critical point of $f: M \to \mathbb{R}$. The *index* of p is the index of the Hessian of f at p, denoted by $\text{Ind}(p)$ or $\text{Ind}_f(p)$.

This number gives us valuable information about the local behavior of f near x. Suppose that $M = \mathbb{R}^n$ and $p = 0$. The second-order Taylor expansion of f at 0 looks like

$$f(x) = f(0) + \tfrac{1}{2} H_0 f(x) + R(x)$$

where $R(x)/|x|^2 \to 0$ as $x \to 0$. Thus f is approximately a constant plus half the Hessian at 0. Let $E_- \oplus E_+$ be a direct sum decomposition of $\mathbb{R}^n$ so that $H_0 f$ is negative definite on E_- and positive definite on E_+, and

$$\dim E_- = k = \operatorname{Ind} H_0 f$$
$$\dim E_+ = n - k$$

Then if $x \in \mathbb{R}^n$ and $t \in \mathbb{R} - 0$ are sufficiently small, $f(tx)/t$ is a decreasing function of t for $x \in E_-$ and an increasing function of t for $x \in E_+$.

It follows that for a Morse function $f: M \to \mathbb{R}$, a critical point p is a local minimum if and only if $\operatorname{Ind}_p = 0$, and a local maximum if and only if $\operatorname{Ind}_p = \dim M$.

The following result of Marston Morse is a sharper form of this relation between f and its Hessian at a nondegenerate critical point p. It states that f has a local representation at p which equals $f(p) + \frac{1}{2}H_p f$.

1.1. Morse's Lemma. *Let $p \in M$ be a nondegenerate critical point of index k of a C^{r+2} map $f: M \to \mathbb{R}$, $1 \leqslant r \leqslant \omega$. Then there is a C^r chart (φ, U) at p such that*

$$f\varphi^{-1}(u_1, \ldots, u_n) = f(p) - \sum_{i=1}^{k} u_i^2 + \sum_{i=k+1}^{n} u_i^2$$

The proof is based on the following parametric form of the diagonalization of symmetric matrices. Let ${}^tQ$ denote the transpose of the matrix Q.

Lemma. *Let $A = \operatorname{diag}\{a_1, \ldots, a_n\}$ be a diagonal $n \times n$ matrix with diagonal entries ± 1. Then there exists neighborhood N of A in the vector space of symmetric $n \times n$ matrices, and a C^ω map*

$$P: N \to GL(n, \mathbb{R})$$

such that $P(A) = I$ (the identity matrix), and if $P(B) = Q$ then ${}^tQBQ = A$.

Proof of Lemma. Suppose B is a symmetric matrix so close to A that b_{11} is nonzero and has the same sign as a_{11}. Consider the linear coordinate change in $\mathbb{R}^n$: $x = Ty$ where

$$x_1 = \left[y_1 - \frac{b_{12}}{b_{11}} y_2 - \cdots - \frac{b_{1n}}{b_{11}} y_n \right] \Big/ \sqrt{|b_{11}|}$$
$$x_k = y_k \qquad \text{for} \qquad k = 2, \ldots, n.$$

One verifies that ${}^tTBT$ has the form

$$\begin{bmatrix} a_{11} & 0 & \cdots & 0 \\ 0 & & & \\ \cdot & & & \\ \cdot & & B_1 & \\ \cdot & & & \\ 0 & & & \end{bmatrix}$$

If B is near enough to A then the symmetric $(n-1) \times (n-1)$ matrix B_1 will be as close as desired to the diagonal matrix $A_1 = \text{diag}\{a_2, \ldots, a_n\}$; in particular it will be invertible. Note that T and B_1 are C^ω functions of B. By induction on n we assume there exists a matrix $Q_1 = P_1(B_1) \in GL(n-1)$ depending analytically on B_1, such that ${}^tQ_1 B_1 Q = A_1$. Define $P(B) = Q$ by $Q = TS$ where

$$S = \begin{bmatrix} 1 & 0 & \cdots & 0 \\ 0 & & & \\ \cdot & & & \\ \cdot & & Q_1 & \\ \cdot & & & \\ 0 & & & \end{bmatrix};$$

then ${}^tQBQ = A$.

QED

Proof of Morse's Lemma. We may assume M is a convex open set in $\mathbb{R}^n$, $p = 0 \in \mathbb{R}^n$, and $f(0) = 0 \in \mathbb{R}$. By a linear coordinate change we may assume that the matrix

$$A = \left[\frac{\partial^2 f}{\partial x_i\, \partial x_j}(0)\right]$$

is diagonal, with the first k diagonal entries equal to -1 and the rest equal to $+1$. By assumption $Df(0) = 0$.

There exists a C^r map $x \mapsto B_x$ from M to the space symmetrix $n \times n$ matrices such that if $x \in M$ and $B_x = [b_{ij}(x)]$ then

$$f(x) = \sum_{i,j=1}^{n} b_{ij}(x) x_i x_j$$

and $B_0 = A$. This follows, for example, from the fundamental theorem of calculus applied twice:

$$\begin{aligned} f(x) &= \int_0^1 Df(tx)x\, dt \\ &= \sum_{j=1}^{n} \left[\int_0^1 \frac{\partial f}{\partial x_j}(tx)\, dt\right] x_j \\ &= \sum_{i,j=1}^{n} \left[\int_0^1 \int_0^1 \frac{\partial^2 f}{\partial x_j\, \partial x_i}(stx)\, ds\, dt\right] x_i x_j \\ &= \sum_{i,j=1}^{n} b_{ij}(x) x_i x_j. \end{aligned}$$

Let $P(B)$ be the matrix valued function in the lemma; put $P(B_x) = Q_x \in GL(n)$. Define a C^r map $\varphi: U \to \mathbb{R}^n$, where $U \subset M$ is a sufficiently small neighborhood of 0.

$$\varphi(x) = Q_x^{-1} x.$$

A calculation shows that $D\varphi(0) = I$; therefore by the inverse function theorem we may assume (φ,U) is a C^r chart.

Put $y = \varphi(x)$; then, in matrix notation:

$$\begin{aligned} f(x) &= {}^t x B_x x \\ &= {}^t y({}^t Q_x B_x Q_x) y \\ &= {}^t y A y \\ &= \sum_{i=1}^{n} a_{ii} y_i^2. \end{aligned}$$

QED

We now have a *complete local description* of a Morse function $f: M \to \mathbb{R}$. If $a \in M$ is a regular point then by the implicit function theorem there are coordinates near a such that

$$f(x_1, \ldots, x_n) = x_1.$$

If a is a critical point there are coordinates near a such that

$$f(x_1, \ldots, x_n) = f(a) - x_1^2 - \cdots - x_k^2 + x_{k+1}^2 + \cdots + x_n^2.$$

The index k is uniquely determined by the critical point.

It follows that the level sets $f^{-1}(y)$ of a Morse function have nice local structure. Near a regular point $f^{-1}(y)$ looks like a hyperplane in $\mathbb{R}^n$. At a critical point there is a chart (φ,U) throwing $U \cap f^{-1}(y)$ onto a neighborhood of 0 in the degenerate quadric hypersurface

$$-x_1^2 - \cdots - x_k^2 + x_{k+1}^2 + \cdots + x_n^2 = 0;$$

nearby level surfaces in U go onto open subsets of the nondegenerate quadrics

$$-x_1^2 - \cdots - x_k^2 + x_{k+1}^2 + \cdots + x_n^2 = \text{constant} \neq 0.$$

See Figure 6–1 for some examples.

As the value of the Morse function increases past a critical value, the topological character of the level surfaces changes suddenly. This is studied in detail in the following sections.

We close this section with:

1.2. Theorem. *For any manifold M, Morse functions from a dense open set in $C_S^s(M,\mathbb{R})$, $2 \leqslant s \leqslant \infty$.*

Proof. The cotangent vector bundle T^*M is isomorphic to $J^1(M,\mathbb{R})$, the bundle of 1-jets of maps $M \to \mathbb{R}$; a natural isomorphism is defined by sending $j_x^1 f \in J_x^1(M,\mathbb{R})$ to $Df_x \in T_x^*M$. Thus a C^s map $f: M \to \mathbb{R}$ is a Morse function if and only if its 1-prolongation

$$j^1 f: M \to J^1(M,\mathbb{R})$$

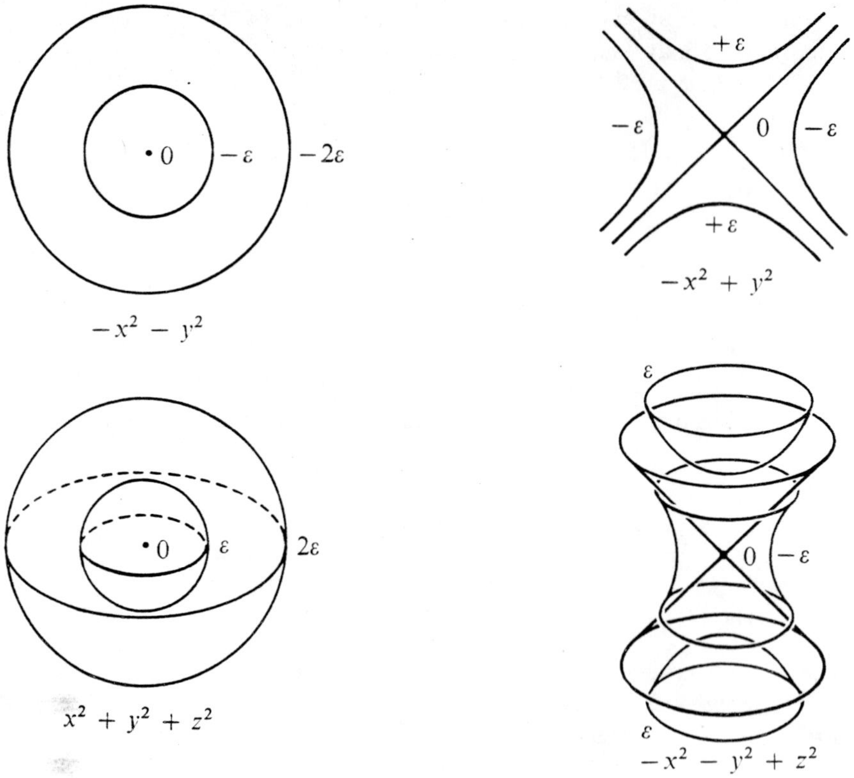

Figure 6–1.

is transverse to the zero section. The theorem now follows from jet transversality (Theorem 3.2.7).

QED

Exercises

****1.** In Morse's lemma C^{r+2} can be replaced by C^{r+1}. (Assume $M = \mathbb{R}^n$, $p = 0$, $f(p) = 0$, Put $\frac{1}{2}D^2f(0)(x,x) = Q(x)$. Let $t \mapsto \xi(t,x)$ be the solution of the differential equation $dx/dt = \text{grad } Q(x)$ such that $\xi(0,x) = 0$. For x near 0 there is a unique $t(x)$ such that $Q\xi(t(x),x) = f(x)$. Define $\varphi(x) = \xi(t(x),x)$. See Kuiper [1], Takens [1].)

2. Let $M \subset \mathbb{R}^{q+1}$ be a compact C^2 submanifold. For each $v \in S^q$ let $f_v: M \to \mathbb{R}$ be the map $f_v(x) = \langle v,x \rangle$. (This is essentially orthogonal projection into the line through v.) Then the set of $v \in S^q$ such that f_v is a Morse function is open and dense.

3. Let $M \subset \mathbb{R}^q$ be a C^2 submanifold and $f: M \to \mathbb{R}$ a C^2 map. The set of linear maps $L \in L(\mathbb{R}^q,\mathbb{R})$ such that the map $M \to \mathbb{R}$, $x \mapsto f(x) + L(x)$ is a Morse function, is a Baire set and thus dense. If M is compact it is open and dense.

4. Let $M \subset \mathbb{R}^q$ be a closed compact C^2 submanifold. The set of points $u \in \mathbb{R}^q$ such that the map $x \mapsto |x - u|^2$ is a Morse function on M, is open and dense.

***5.** Let $f_0, f_1 : M \to \mathbb{R}$ be Morse functions. It may be impossible to find a C^2 homotopy $f: M \times [0,1] \to \mathbb{R}$ such that $f(x,0) = f_0(x)$, $f(x,1) = f_1(x)$ and each map $f_t(x) = f(x,t)$ is a Morse function, $0 \leqslant t \leqslant 1$. (Take $M = \mathbb{R}$, $f_0(x) = x^3 - x$, $f_1(x) = x^3 + x$.) However, a C^2 homotopy from f_0 to f_1 can be found such that each f_t is a Morse function except for a finite set $t_0, \ldots, t_m$; for each j, f_{t_j} has only one degenerate critical point z_j; in suitable local coordinates at z_j, f_{t_j} has the form:

$$-x_1^2 - \cdots - x_k^2 + x_{k+1}^2 + \cdots + x_{n-1}^2 \pm x_n^3 + R(x) + \text{constant}$$

where $R(x)/|x|^3 \to 0$ as $|x| \to 0$. (Assume f is C^3 and make the map $(x,t) \mapsto j_x^2 f_t$ transverse to suitable submanifolds of $J^2(M,\mathbb{R})$.)

6. There is a Morse function on the projective plane which has exactly three critical points.

***7.** *Generalized Morse's lemma.* Let $f: M \to \mathbb{R}$ be a C^{r+3} map. A submanifold $V \subset M$ is *critical* if every point of V is a critical point. A critical submanifold V is *nondegenerate of index k* if every $x \in V$ has the following property: for some (hence any) submanifold $W \subset M$ which is transverse to V at x, the point x is a nondegenerate critical point of $f|W$ having index k. If also $V \subset M - \partial M$, and V is connected, then there is a C^r tubular neighborhood $(g, \xi \oplus \eta)$ for (M,V) and an orthogonal structure on $\xi \oplus \eta$ such that the composition $E(\xi \oplus \eta) \underset{g}{\subset} M \to \mathbb{R}$ is given by

$$(x, y) \to -|x|^2 + |y|^2 + C$$

for $(x, y) \in \xi_p \oplus \eta_p$, $p \in V$, where C is the constant $f(V)$.

2. Differential Equations and Regular Level Surfaces

We recall some facts about differential equations. Let $W \subset \mathbb{R}^n$ be an open set and $g: W \to \mathbb{R}^n$ a C^r map, $1 \leqslant r \leqslant \omega$, regarded as a vector field on W. Then locally g satisfies Lipschitz conditions, so the basic theorems about existence, uniqueness and differentiability of solutions of ordinary differential equations apply to the initial value problem:

$$\varphi'(t) = g(\varphi(t)),\qquad \varphi(0) = x \tag{1}$$

for each $x \in W$. Therefore there is an open interval $J \subset \mathbb{R}$ about 0 and a C^{r+1} map

$$\varphi: (J,0) \to (W,x)$$

which satisfies (1). If $\varphi_1 : J_1 \to W$ is another solution to (1) then $\varphi = \varphi_1$ on $J \cap J_1$. Thus φ and φ_1 fit together to form a solution on $J \cup J_1$. It follows that J and φ are unique provided J is taken to be maximal. We call this maximal interval $J(x)$, and the corresponding solution is written variously as

$$\varphi^x : J(x) \to W,$$
$$\varphi^x(t) = \varphi_t(x) = \varphi(t,x).$$

The maps φ^x and sometimes the sets $\varphi^x(J(x))$ are called *solution curves* or *trajectories* or *flow lines* of the vector field f.

The interval-valued function $x \mapsto J(x)$ is lower semicontinuous in the sense that if $\alpha \in J(x)$ then $\alpha \in J(y)$ for all y in some neighborhood of x. This implies that the set

$$\Omega = \{(t,x) \in \mathbb{R} \times W : t \in J(x)\}$$

is open in $\mathbb{R} \times W$.

The *flow* generated by f is the C^r map

$$\varphi : \Omega \to W,$$
$$(t,x) \mapsto \varphi_t(x).$$

For each $t \in \mathbb{R}$ let

$$W_t = \{x \in W : t \in J(x)\}.$$

Then W_t is open in W and there is defined a C^r map

$$\varphi_t : W_t \to W,$$
$$x \mapsto \varphi_t(x).$$

Clearly $\varphi_t(W_t) = W_{-t}$ and $\varphi_{-t} = \varphi_t^{-1}$. Thus *$\varphi_t$ is a C^r embedding.* More generally we have the relations

$$\varphi_s \varphi_t(x) = \varphi_{s+t}(x),$$

valid in the sense that if one side is defined so is the other, and they are equal. It is not hard to prove that if $K \subset \mathbb{R}$ is any interval and $U \subset W$ is an open set such that $U \subset \bigcap_{t \in K} W_t$, then the map $K \to \mathrm{Emb}^r(K,W)$ is continuous for the weak topology.

Let $P \subset W$ be compact. For each $x \in W$ the set of $t \in J(x)$ such that $\varphi_t(x) \in P$, is closed in $\mathbb{R}$, not merely in $J(x)$. This has the important consequence that if $x \in P$ is such that $\varphi_t(x) \in P$ for all $t \in J(x) \cap \mathbb{R}_+$ then $J(x) \supset \mathbb{R}_+$, and similarly for $\mathbb{R}_-$. In particular *if the trajectory of x has compact closure in W then $J(x) = \mathbb{R}$.*

Now let X be a C^r vector field on an n-manifold M; that is, X is a C^r section of TM. Assume first that $\partial M = \varnothing$.

An *integral curve* (or *solution curve*) of X is a differentiable map $\eta : J \to M$ where $J \subset \mathbb{R}$ is an interval and $\eta'(t) = X(\eta(t))$ for each $t \in J$. If (ψ, U) is a chart on M (of class C^∞, or C^ω if $r = \omega$) containing $\eta(J)$ and $W = \psi(U) \subset \mathbb{R}^n$ then the composite map

$$f : W \xrightarrow{\psi^{-1}} U \xrightarrow{X} TU \xrightarrow{D\psi} \mathbb{R}^n$$

is a C^r vector field on W. The map $\varphi = \psi \circ \eta : J \to W$ satisfies the differential equation

$$\varphi'(t) = f(\varphi(t)), \tag{2}$$

because

$$\begin{aligned} \varphi'(t) &= D\psi(\eta'(t)) = D\psi(X_\eta(t)) \\ &= (D\psi \circ X \circ \psi^{-1})(\psi\eta(t)) \\ &= f(\varphi(t)). \end{aligned}$$

Thus ψ carries an integral curve of X into solution of (2), provided the integral curve lies in a coordinate domain.

All the results about vector fields on open sets in $\mathbb{R}^n$ carry over to vector fields on M. For each $x \in M$ there is a maximal open interval $J(x)$ about 0 and a C^{r+1} integral curve (or *trajectory*, or *flow line*) of X,

$$\eta^x:(J(x),0) \to (M,x)$$
$$\eta^x(t) = \eta(t,x) = \eta_t(x).$$

The set

$$\Omega = \{(t,x) \in \mathbb{R} \times M : t \in J(x)\}$$

is open in $\mathbb{R} \times M$; the *flow* of X is the C^r map

$$\eta:\Omega \to M,$$
$$(t,x) \mapsto \eta_t(x).$$

The previous results about endpoints of $J(x)$ and compact subsets of M are still valid. An important case is where M is compact and without boundary. In this case $\Omega = M \times \mathbb{R}$ and each η_t is a C^r diffeomorphism of M. Thus the maps $\{\eta_t\}_{t \in \mathbb{R}}$ form a *one-parameter group* of C^r diffeomorphism of M: the map

$$\mathbb{R} \to \quad \mathrm{Diff}^r(M), \qquad t \mapsto \eta_t$$

is a continuous homomorphism, and $\eta_t\eta_s = \eta_{t+s}$.

Unfortunately we must also consider vector fields on ∂-manifolds. Suppose now that $\partial M \neq \varnothing$. The preceding results can be used if we first embed M as a closed submanifold of an n-manifold N without boundary, such as the double of M, and then extend X to a C^r vector field on N. (This can be done with a partition of unity if $1 \leqslant r \leqslant \infty$, since local C^r extensions exist by definition. If $r = \omega$ the local extensions are unique and fit together to give an analytic vector field on a neighborhood of M in N, which is all that is needed.)

If X is tangent to ∂M, that is, if $X(\partial M) \subset T(\partial M)$, everything is as before. But if X is not tangent to ∂M, the intervals $J(x)$ will not all be open. If $x \in \partial M$, $X(x) \neq 0$, and $X(x)$ points into [respectively, out of] M, then $J(x)$ will contain 0 as its left [resp. right] endpoint. It is also possible that $X(x)$ is tangent to ∂M and still $J(x)$ contains the endpoint 0; and $J(x)$ can even be the degenerate interval $\{0\}$. For any $y \in M$, if $J(y)$ contains an endpoint b then $\eta(b,y) \in \partial M$.

The set Ω, defined as before, is not necessarily open in $\mathbb{R} \times M$, but its interior is dense. Moreover the flow $\eta:\Omega \to M$ is C^r in the sense that it extends to a C^r map $\Omega' \to N$ where $\Omega' \subset \mathbb{R} \times N$ is open.

If the trajectory of $x \in M$ has compact closure then $J(x)$ is a closed interval; if also the trajectory lies in $M - \partial M$ then $J(x) = \mathbb{R}$.

If $J(x) = \mathbb{R}$ for all $x \in M$ the vector field is called *completely integrable*. A necessary condition for X to be completely integrable is that X be tangent to ∂M. A sufficient condition is that X be tangent to ∂M and each trajectory have compact closure. A more general sufficient condition is given in Exercise 1.

Differential equations can be used to prove anew the *collaring theorem*:

2.1. Theorem. *Let M be a ∂-manifold. Then there exists a C^∞ embedding*

$$F:\partial M \times [0,\infty) \to M$$

such that $F(x,0) = x$ for all $x \in \partial M$.

Proof. Using charts covering ∂M and a partition of unity, one finds a C^∞ vector field X on a neighborhood $U \subset M$ of ∂M which is nowhere tangent to ∂M and which points into M (in local coordinates). Let $W \subset \partial M \times [0,\infty)$ be a neighborhood of $\partial M \times 0$ on which the flow η of the vector field is defined. There is a C^∞ embedding $h:\partial M \times [0,\infty) \to W$ which leaves $\partial M \times 0$ pointwise fixed. The required map F is the composition

$$F:\partial M \times [0,\infty) \xrightarrow{h} W \xrightarrow{\eta} M.$$

QED

We turn now to the construction of a vector field transverse to the regular level surfaces of a C^{r+1} map $f:M \to R$, $r \geqslant 1$. We assume M has been given a C^∞ Riemannian metric. The inner product in any M_x is denoted by $\langle X,Y\rangle$; the corresponding norm is $|X| = \langle X,X\rangle^{1/2}$.

For every linear map $\lambda:M_x \to \mathbb{R}$ there exists a unique tangent vector $X_\lambda \in M_x$ such that $\lambda(Y) = \langle X_\lambda,Y\rangle$ for all $Y \in M_x$. We call X_λ the vector *dual to* λ. The map $\lambda \mapsto X_\lambda$ is a linear isomorphism from M_x^* onto M_x. Its inverse assigns to $X \in M_x$ the linear map

$$\lambda_X:M_x \to \mathbb{R}, \qquad Y \mapsto \langle X,Y\rangle.$$

If $f:M \to \mathbb{R}$ is C^{r+1}, for each $x \in M$ define the vector grad $f(x) \in M_x$ to be the dual of Df_x. In this way the C^r *gradient vector field* grad f is defined. It depends on the Riemannian metric.

If M is open in $\mathbb{R}^n$ and the metric is given by the standard inner product of $\mathbb{R}^n$ then

$$\operatorname{grad} f(x) = \left(\frac{\partial f}{\partial x_1}(x), \ldots, \frac{\partial f}{\partial x_n}(x)\right).$$

It is clear that grad $f(x) = 0$ if and only if x is a critical point of f. At a regular point grad $f(x)$ is transverse to the level surface $f^{-1}(f(x))$; in fact, they are orthogonal.

Notice that f is nondecreasing along gradient lines, that is, along solution curves of the gradient differential equation $\eta' = \operatorname{grad} f(\eta)$. For if $\eta(t)$ is a solution then

$$\frac{d}{dt} f(\eta(t)) = \langle \operatorname{grad} f(\eta(t)), \operatorname{grad} f(\eta(t))\rangle$$

$$= |\operatorname{grad} f(\eta(t))|^2 \geqslant 0.$$

And f is strictly increasing along any solution curve which is not a critical point.

The following *regular interval theorem* is a useful way of finding diffeomorphisms.

2.2. Theorem. *Let $f: M \to [a,b]$ be a C^{r+1} map on a compact ∂-manifold, $1 \leqslant r \leqslant \omega$. Suppose f has no critical points and $f(\partial M) = \{a,b\}$. Then there is a C^r diffeomorphism $F: f^{-1}(a) \times [a,b] \to M$ so that the diagram*

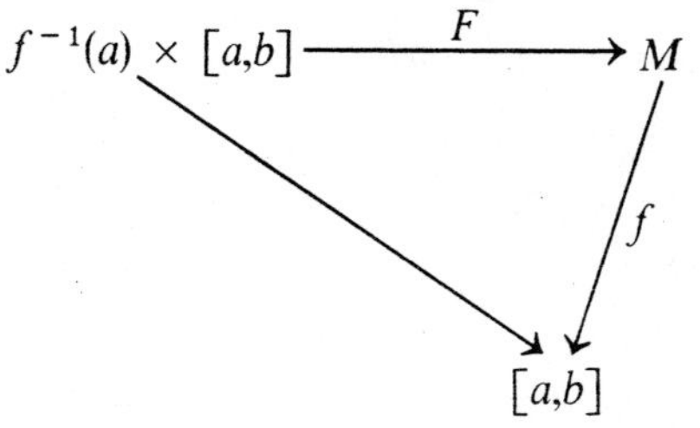

commutes. In particular all level surfaces of f are diffeomorphic.

Proof. Give M a Riemannian metric. Consider the C^r vector field on M:

$$X(x) = \frac{\operatorname{grad} f(x)}{|\operatorname{grad} f(x)|^2}.$$

Notice that $X(x)$ has the same trajectories as grad f but with a different parametrization.

Let $\eta: [t_0, t_1] \to M$ be a solution curve of X. A computation shows that the derivative of the map

$$[t_0, t_1] \to \mathbb{R}, \qquad t \mapsto f(\lambda(t))$$

is identically 1. This means that

(1) $$f\eta(t_1) - f\eta(t_0) = t_1 - t_0.$$

Let $x \in f^{-1}(s)$. Since M is compact, the set $J(x)$ is closed; from (1) it follows that

(2) $$J(x) = [a - s, b - s].$$

The assumptions on f imply that $f^{-1}(a)$ is a union of boundary components of M. Define a map

$$F: f^{-1}(a) \times [a,b] \to M,$$
$$F(x,t) = \eta(t - a, x).$$

Since f increases along gradient lines, and thus along X-trajectories, F is injective. And F is an immersion because gradient lines are transverse to level surfaces. Thus F is an embedding. Finally, F is onto because of (2).

QED

2.3. Corollary. *Let M be a compact manifold and assume $\partial M = A \cup B$ where A and B are disjoint closed sets. Suppose there exists a C^2 map $f: M \to \mathbb{R}$ without critical point such that $f(A) = 0$, $f(B) = 1$. Then M is diffeomorphic to both $A \times I$ and $B \times I$.*

The following topological application of critical point theory is due to G. Reeb.

2.4. Theorem. *Let M be a compact n-dimensional manifold without boundary, admitting a Morse function $f: M \to \mathbb{R}$ with only 2 critical points. Then M is homeomorphic to the n-sphere S^n.*

Proof. Let the critical points be P_+ and P_-. We may assume P_+ is a maximum and P_- a minimum. Put $f(P_+) = z_+$, $f(P_-) = z_-$. By Morse's lemma there are coordinates $(x_1, \ldots, x_n)$ in a neighborhood U_+ of P_+ giving $f|U_+$ the form

$$f(P_+) = -x_1^2 - \cdots - x_n^2 + z_+.$$

Therefore there exists $b < z_+$ such that the set

$$D_+ = f^{-1}[b, z_+]$$

is a neighborhood of P_+ diffeomorphic to the n-disk D^n.

Similarly there exists $a > z_-$ such that the set

$$D_- = f^{-1}[z_-, a]$$

is a neighborhood of P_- diffeomorphic to D^n. We assume $z_- < a < b < z_+$. Note that

$$\partial D_+ \approx \partial D_- \approx S^{n-1}.$$

By Theorem 2.2 the set $f^{-1}[a,b]$ is diffeomorphic to $S^{n-1} \times I$. See Figure 6–2.

Let $Q_+, Q_- \subset S^n$ be the north and south poles. Let B_+, B_- be disjoint neighborhoods of Q_+, Q_- diffeomorphic to D^n (the two "polar caps") so that, putting $C = S^n - \text{Int}(B_+ \cup B_-)$, we have $C \approx S^{n-1} \times I$ and $\partial C = \partial B_+ \cup \partial B_-$.

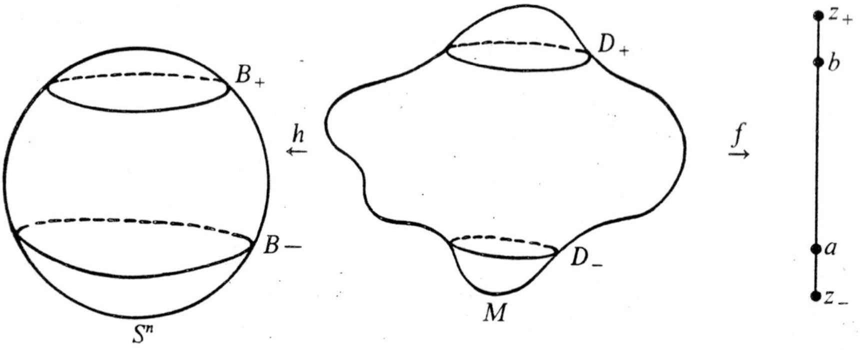

Figure 6–2.

Let $h_0: D_+ \to B_+$ be a diffeomorphism. Extend

$$h_0 | \partial D_+ : \partial D_+ \to \partial B_+$$

to a diffeomorphism $(\partial D_+) \times I \to (\partial B_+) \times I$. This provides an extension of h_0 to a homeomorphism

$$h_1 : D_+ \cup f^{-1}a, b \to B_+ \cup C.$$

It is possible to extend

$$h_1 | \partial D_- : \partial D_- \to \partial B_-$$

to a homeomorphism $D_- \to B_1$. This is the same as extending a homeomorphism $g_0 : S^{n-1} \to S^{n-1}$ to a homeomorphism $g: D^n \to D^n$, and one can extend *radially*:

$$g(x) = \begin{cases} |x| g_0(x/|x|) & \text{if } x \neq 0 \\ 0 & \text{if } x = 0. \end{cases}$$

Thus g maps each radial segment $[0,y]$, $y \in S^{n-1}$, linearly onto the segment $[0, g_0(y)]$.

In this way h_1 is extended to a homeomorphism $h: M \to S^n$.

QED

It is not always possible to find a *diffeomorphism* between M and S^n! In 1956 John Milnor [1] found an example of a manifold which is homeomorphic, but not diffeomorphic, to S^7. This very surprising result stimulated intensive research into such "exotic spheres" and into the more general problem of finding and classifying all differential structures on a manifold. A great deal is now known, but the problem has not been solved.

Exercises

1. A vector field X on M has *bounded velocity* if there is a complete Riemannian metric on M such that $|X(x)|$ is bounded. In this case every maximal solution curve is defined on a closed interval $J(x)$. If also X is tangent to ∂M then X is completely integrable.

*****2.** If a C^1 vector field X is completely integrable, does X necessarily have bounded velocity? (See Exercise 1.)

3. Let X be a C^r vector field on a manifold M, $1 \leqslant r \leqslant \infty$. There is a completely integrable C^r vector field Y on M whose trajectories (considered as subsets) are the same as those of X.

4. Let X be a C^1 vector field on a ∂-manifold M. If $x \in \partial M$ and $J(x) = [0,a]$ or $[0,a)$ then every neighborhood of x contains a point $y \in \partial M$ such that $X(y) \pitchfork \partial M$ and $X(y)$ points inward.

5. Let (x, y) be local coordinates on the torus $T = S^1 \times S^1$ corresponding to angular variables on S^1 taken mod 2π. For each pair α, β of real numbers not both zero, let $X_{\alpha,\beta}$ denote the vector field on T which in (x, y) coordinates is the constant field (α, β).

(a) If α and β are linearly dependent over the rational numbers then every trajectory of $X_{\alpha,\beta}$ is a circle.

(b) If α and β are linearly independent over the rationals then every trajectory of $X_{\alpha,\beta}$ is dense. In fact:

(c) In case (b), let τ be any nonzero real number. Then for each $x \in M$ the set $\{\varphi_{n\tau}(x): n \in \mathbb{Z}_+\}$ is dense, where φ_t is the flow of $X_{\alpha,\beta}$.

*6. On every surface of genus $p \geqslant 1$ there is a vector field having a dense trajectory.

***7. Does every C^2 vector field on S^3 necessarily have either a zero or a periodic trajectory? (This has recently been proved false for C^1 fields by Paul Schweitzer [1].)

*8. Let X be a completely integrable C^1 vector field on a manifold M. Suppose that every *positive semi-orbit* $\{\varphi_t(x): t \geqslant 0\}$ is dense and that every *negative semi-orbit* $\{\varphi_t(x): t \leqslant 0\}$ is dense, where φ is the flow generated by X. Then M is compact.

9. Let $f: M \to \mathbb{R}$ be a C^1 function on a Riemannian manifold and let $V \subset M$ be a submanifold. If $x \in V$ then $\operatorname{grad}(f|V)(x)$ is the image of $\operatorname{grad} f(x)$ under the orthogonal projection $M_x \to V_x$.

10. Theorem 2.2 is true for noncompact M under the extra hypothesis that M has a complete Riemannian metric for which $|\operatorname{grad} f(x)|$ is bounded. (See Exercise 1). But without this assumption there are simple counter examples.

11. The trick used in the proof of Theorem 2.2, of following integral curves, can often be used to obtain diffeomorphisms. For example, let X be a C^r vector field on M, $1 \leqslant r \leqslant \omega$. Let V_0, V_1 be C^r submanifolds of M which are transverse to X. Assume $\partial V_0 = \partial V_1 = \partial M = \varnothing$. Suppose that every integral curve through a point of either of the submanifolds intersects the other at a unique point. Then V_0 and V_1 are C^r diffeomorphic. Moreover if $r < \omega$ there is a C^r diffeomorphism of M which is C^r isotopic to the identity and which carries V_0 onto V_1.

12. Corollary 2.3 is also true for C^1 maps.

13. Theorem 2.4 admits the stronger conclusion that M is the union of two n-balls intersecting along their common boundary.

*14. Theorem 2.4 can be generalized as follows. If M admits a function with only two critical points (perhaps degenerate), then the complement of either critical point is diffeomorphic to $\mathbb{R}^n$ and M is homeomorphic to S^n. (Use the gradient flow and a disk around a critical point P to exhibit $M - P$ as an increasing union of open disks; then use Exercise 15 of Section 1.2.)

***15. The conclusions of Exercises 13 and 14 suggest the difficult problem: if $M - P \approx \mathbb{R}^n$, is M the union of two n-balls intersecting in their common boundary? This is known to be true for all $n \neq 4$.

3. Passing Critical Levels and Attaching Cells

In order to make ∂M behave nicely with respect to level surfaces, we shall consider Morse functions $f: M \to [a,b]$ of the following type, which we call *admissible*: $\partial M = f^{-1}(a) \cup f^{-1}(b)$, and a and b are regular values. This has the following implications. Each of $f^{-1}(a)$, $f^{-1}(b)$ is a union of components of ∂M. If $f^{-1}(a)$ or $f^{-1}(b)$ is empty, the minimum or, respectively, maximum value of f is taken only at critical points in $M - \partial M$.

Theorem 1.2 implies that a compact manifold has an admissible Morse function taking prescribed constant values on boundary components.

Figure 6–3 shows some examples of admissible Morse functions on 2-manifolds, the map being orthogonal projection into the vertical interval $[a,b]$. Note that if $p, q \in [a,b]$ are regular values, $p < q$, then restriction of f to $f^{-1}[p,q]$ is also admissible (into $[p,q]$).

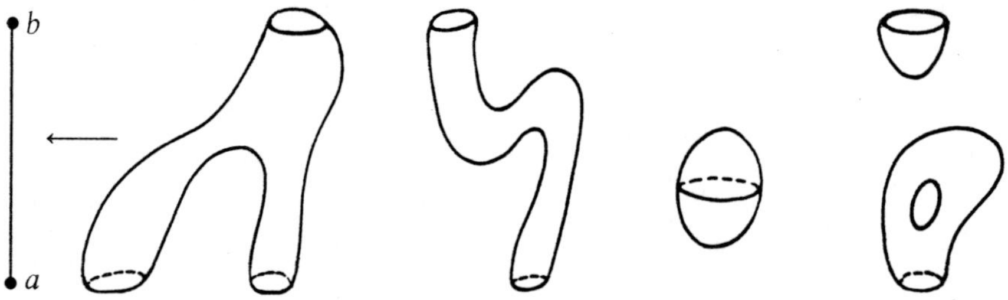

Figure 6–3. Admissible Morse functions.

In the last section we saw that if f has no critical points then $M \approx f^{-1}(a) \times I$. In the following theorem we suppose f has just one critical point.

By a *k-cell* in M is meant the image of an embedding $D^k \hookrightarrow M$.

3.1. Theorem. *Let M be compact and $f: M \to [a,b]$ an admissible Morse function. Suppose f has a unique critical point z, of index k. Then there exists a k-cell $e^k \subset M - f^{-1}(b)$ such that $e^k \cap f^{-1}(a) = \partial e^k$, and there is a deformation retraction of M onto $f^{-1}(a) \cup e^k$.*

Proof. Let $f(z) = c$, $a < c < b$. To prove the theorem it suffices to prove it for the restriction of f to $f^{-1}[a',b']$ for any a', b' such that $a < a' < c < b' < b$, by the regular interval Theorem 2.2 applied to $f^{-1}[a,a']$ and $f^{-1}[b,b']$. Moreover, we can assume $c = 0$, replacing f by $f(x) - c$ otherwise.

Let (φ, U) be a chart at z as in Morse's lemma. We write $\mathbb{R}^n = \mathbb{R}^k \times \mathbb{R}^{n-k}$; thus φ maps U diffeomorphically onto an open set $V \subset \mathbb{R}^k \times \mathbb{R}^{n-k}$, and

$$f\varphi^{-1}(x,y) = -|x|^2 + |y|^2$$

for $(x,y) \in V$. Note that $\varphi(z) = (0,0)$. Put $g(x,y) = -|x|^2 + |y|^2$.

Let $0 < \delta < 1$ be such that V contains $\Gamma = B^k(\delta) \times B^{n-k}(\delta)$ where $B^i(\delta) \subset \mathbb{R}^i$ is the closed ball about 0 of radius δ. Give M a Riemannian metric which agrees in $\varphi^{-1}(\Gamma)$ with the metric induced by φ from the standard inner product on $\mathbb{R}^n$. If $\varphi(u) = v \in \Gamma$ then

$$D\varphi(u)(\operatorname{grad} f(u)) = \operatorname{grad} g(v).$$

Let $\varepsilon > 0$ be much smaller than δ, say $\varepsilon < \delta^2/100$. Put

$$\begin{aligned} B^k &= B^k(\sqrt{\varepsilon}) \times 0 \subset \mathbb{R}^k \times \mathbb{R}^{n-k} \\ &= \{(x,0) \in \mathbb{R}^k \times \mathbb{R}^{n-k} : |x|^2 \leqslant \varepsilon\}, \end{aligned}$$

and $e^k = \varphi^{-1}(B^k)$.

A deformation of $f^{-1}[-\varepsilon,\varepsilon]$ to $f^{-1}(-\varepsilon) \cup e^k$ is made by patching together two deformations. First consider the set

$$\Gamma_1 = D^k(\sqrt{2\varepsilon}) \times D^{n-k}(\sqrt{2\varepsilon}).$$

See Figure 6–4 for the case $k = 1, n = 2$. In $\Gamma_1 \cap g^{-1}[-\varepsilon,\varepsilon]$ a deformation

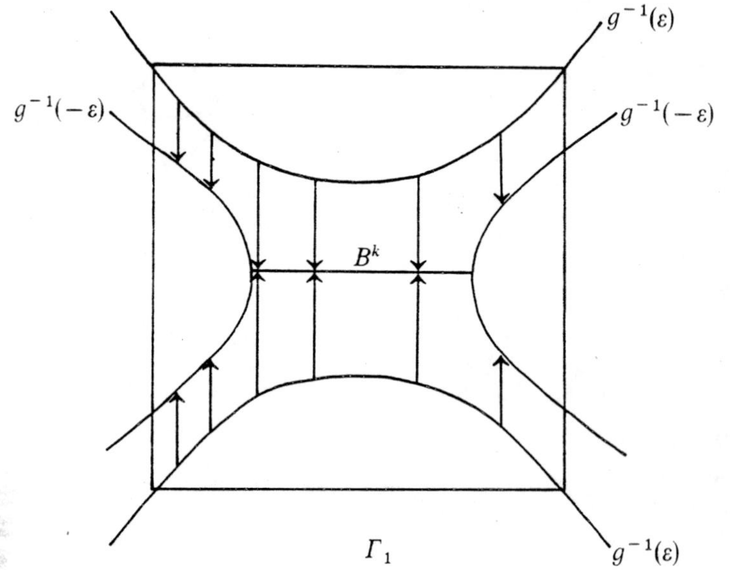

Figure 6–4.

is obtained by moving (x,y) at constant speed along the interval joining (x,y) to the point $(x,sy) \in g^{-1}(-\varepsilon) \cup B^k$, $s \in \mathbb{R}$ where

$$s = s(x,y) = \begin{cases} 0 & \text{if} \quad |x|^2 < \varepsilon \\ \sqrt{|x|^2 - \varepsilon}/|y| & \text{if} \quad |x| \geqslant \varepsilon. \end{cases}$$

Note that these intervals are closures of solution curves of the vector field $X(x,y) = (0.-2y)$. This deformation is transported to $\varphi^{-1}(\Gamma_1)$ via conjugation by φ.

Outside the set

$$\Gamma_2 = D^k(\sqrt{2\varepsilon}) \times D^{n-k}(\sqrt{2\varepsilon})$$

the deformation moves each point at constant speed along the flow line of the vector field $-\text{grad } g$ so that it reaches $g^{-1}(-\varepsilon) \cup B^k$ in unit time. (The speed of each point is the length of its path under the deformation.) See Figure 6–5. This deformation is transported to $U - \varphi^{-1}(\Gamma_2)$ by φ; it is then extended over $M - \varphi^{-1}(\Gamma_2)$ by following flow lines of $-\text{grad } f$. Each such flow line must eventually reach $f^{-1}(-\varepsilon)$, for it can never enter Γ_2 because $|x|$ increases and $|y|$ decreases along flow lines, and $|\text{grad } f|$

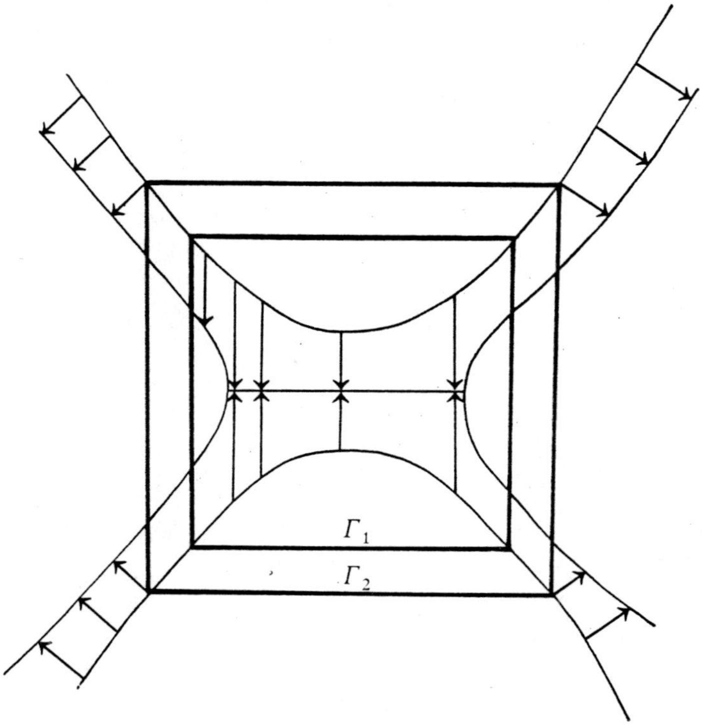

Figure 6–5.

has a positive lower bound in the compact set

$$f^{-1}[-\varepsilon,\varepsilon] - \operatorname{Int} \varphi^{-1}\Gamma_2.$$

To extend the deformation to points of $\Gamma_2 - \Gamma_1$ it suffices to find a vector field on Γ which agrees with X in Γ_1 and with $-\operatorname{grad} g$ in $\Gamma - \Gamma_2$. Such a field is

$$Y(x,y) = 2(\mu(x,y)x, -y)$$

where the C^∞ map $\mu: \mathbb{R}^k \times \mathbb{R}^{n-k} \to [0,1]$ vanishes in Γ_1 and equals 1 outside Γ_2. It is easy to see that each flow line of Y which starts at a point of

$$(\Gamma_2 - \Gamma_1) \cap g^{-1}[-\varepsilon,\varepsilon]$$

must reach $g^{-1}(-\varepsilon)$ because $|x|$ is non-decreasing along flow lines.

The global deformation of $f^{-1}[-\varepsilon,\varepsilon]$ into $f^{-1}(-\varepsilon) \cup e^k$ is obtained by moving each point of Γ at constant speed along the flow line of Y until it reaches $g^{-1}(-\varepsilon) \cup B^k$ in unit time and transporting this motion to M via φ; while each point of $M - \varphi^{-1}(\Gamma)$ moves at constant speed along the flow line of $-\operatorname{grad} f$ until it reaches $f^{-1}(-\varepsilon)$ in unit time. Of course points on $f^{-1}(-\varepsilon) \cup e^k$ stay fixed.

QED

If we consider the map $-f: M \to [-b, -a]$ instead of f, we obtain the following result dual to Theorem 3.1:

3.2. Theorem. *Let $f: M \to [a,b]$ be an admissible Morse function having a unique critical point z, of index k. Then there exists an $(n-k)$-cell $e_*^{n-k} \subset M - f^{-1}(a)$ such that $e_*^{n-k} \cap f^{-1}(b) = \partial e_*^{n-k}$, and there is a deformation retraction of M onto $f^{-1}(b) \cup e_*^{n-k}$.* Moreover *e_*^{n-k} can be chosen so that e^k (of Theorem 3.1) meets e_*^{n-k} only at z, and transversely.*

We speak of these cells e^k and e_*^{n-k} as dual to each other. In Figure 6–6 dual pairs of cells are shown.

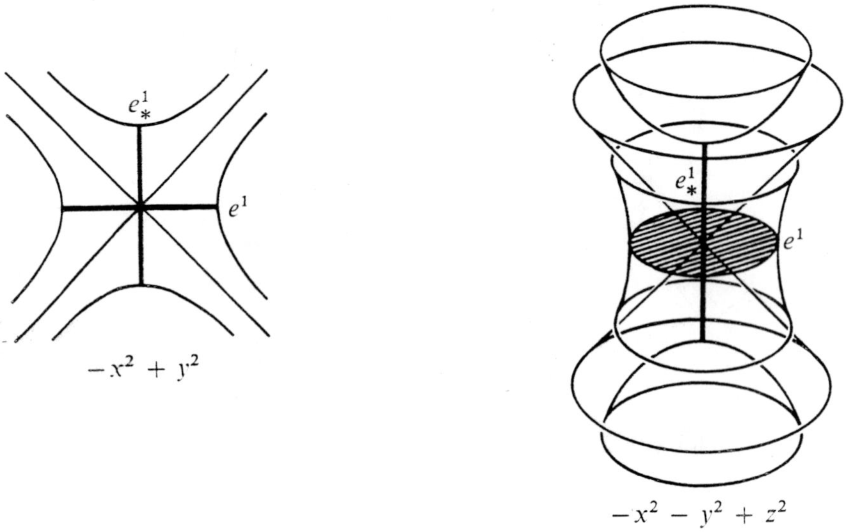

Figure 6–6. Dual cells.

The k'th *type number* of a Morse function $f: M \to \mathbb{R}$ is the number $\nu_k = \nu_k(f)$ of critical points of index k, $0 \leqslant k \leqslant n = \dim M$. We say f has *type* $(\nu_0, \ldots, \nu_n)$.

3.3. Theorem. *Let $f: M \to [a,b]$ be an admissible Morse function of type $(\nu_0, \ldots, \nu_n)$ on a compact manifold. Suppose f has just one critical value c, $a < c < b$. Then there are disjoint k-cells $e_i^k \subset M - f^{-1}(b)$, $1 \leqslant i \leqslant \nu_k$, $k = 0, \ldots, n$, such that $e_i^k \cap f^{-1}(a) = \partial e_i^k$; and there is a deformation retraction of M onto $f^{-1}(a) \cup \{\bigcup_{i,k} e_i^k\}$.*

The proof is just like that of Theorem 3.1, using disjoint Morse charts for the different critical points.

We can now prove the celebrated *Morse inequalities.* These require familiarity with singular homology theory, or any homology theory satisfying the Eilenberg–Steenrod axioms.

The k'th singular homology group of the pair (X,A) with coefficients in the field F is denoted by $H_k(X,A;F)$; this is a vector space over F and thus has a dimension $\lambda_k(X,A;F)$. When F is the rational field Q, these are called the *Betti numbers* of (X,A). If these numbers are finite, and only finitely many are nonzero, then the *homological Euler characteristic*

$$\chi'(X,A) = \sum_{k=0}^{\infty} (-1)^k \lambda_k(X,A;F)$$

of (X,A) is defined. When X is a compact manifold and A is a compact submanifold then $\chi'(X,A)$ is defined and is independent of the field F.

3.4. Theorem. *Let $f:M \to [a,b]$ be an admissible Morse function on a compact manifold, of type $(\nu_0,\ldots,\nu_n)$. Let F be a field and denote by β_k the dimension of the relative homology group $H_k(M,f^{-1}(a);F)$. Then:*

$$\text{(a)} \quad \sum_{k=0}^{m} (-1)^{k+m}\nu_k \geqslant \sum_{k=0}^{m} (-1)^{k+m}\beta_k$$

for $0 \leqslant m \leqslant n$; and

$$\text{(b)} \quad \sum_{k=0}^{n} (-1)^k \nu_k = \sum_{k=0}^{n} (-1)^k \beta_k = \chi'(M,f^{-1}(a)).$$

Before giving the proof we derive some corollaries.

3.5. Theorem. *Let $f:M \to [a,b]$ be an admissible Morse function on a compact manifold, of type $(\nu_0,\ldots,\nu_n)$. Assume that $f^{-1}(a) = \varnothing$. If β_k denotes the dimension of $H_k(M;F)$ where F is a field then (a) and (b) of 3.4 are valid. In particular the alternating sum of the type numbers equals the homological Euler characteristic $\chi'(M) = \sum_{k=0}^{m} (-1)^k\beta_k$.*

Notice that $\nu_k(f) = \nu_{n-k}(-f)$. From this follows:

3.6. Theorem. *The homological Euler characteristic of a compact odd dimensional manifold without boundary is 0.*

Proof. Let $f:M \to \mathbb{R}$ be a Morse function of type $(\nu_0,\ldots,\nu_n)$. Let $a < f(x) < b$ for all $x \in M$; thus $f:M \to [a,b]$ is an admissible Morse function. Then Theorem 3.5 applied to f and $-f$ gives

$$\begin{aligned}\chi'(M) &= \sum_{k=0}^{n} (-1)^k \nu_k = \sum_{k=0}^{n} (-1)^k \nu_{n-k}(-f) \\ &= (-1)^n \sum_{k=0}^{n} (-1)^{n-k} \nu_{n-k}(-f) \\ &= -\chi'(M).\end{aligned}$$

QED

The inequalities (a) in Theorem 3.4 can be conveniently arranged as follows. Put $\nu_k - \beta_k = \delta_k$. Then

$$\delta_0 \geqslant 0$$
$$\delta_1 \geqslant \delta_0 \geqslant 0$$
$$\delta_3 \geqslant \delta_1 - \delta_0 \geqslant 0$$

and in general

$$\delta_{m+1} \geqslant \delta_m - \delta_{m-1} + \cdots + (-1)^m \delta_0 \geqslant 0.$$

In particular, $\delta_k \geqslant 0$, $k = 0, \ldots, n$. This proves:

3.7. Theorem. *In Theorems 3.4 and 3.5,* $\nu_k \geqslant \beta_k$, $k = 0, \ldots, n$.

We now prove Theorem 3.4. It is convenient to assume that f separates the critical points, that is, $f(z_1) \neq f(z_2)$ if z_1, z_2 are distinct critical points. This can be arranged by perturbing f slightly in disjoint neighborhoods $U_i \subset M - \partial M$ of the critical points z_i to get a function $g: M \to \mathbb{R}$ of the following type. Outside the union of the U_i, $g = f$. In U_i we have

$$g(x) = f(x) + \varepsilon_i \lambda_i(x)$$

where the C^∞ map $\lambda_i: M \to [0,1]$ has support in U_i and equals 1 on a neighborhood of z_i, while $\varepsilon_i > 0$. As max ε_i tends to 0, g tends to f in $C^2_S(M,\mathbb{R})$. Thus for small ε_i, g will be a Morse function having the same critical points as f_1 and these will have the same indices. We can choose the ε_i so that g separates the points z_i. Therefore we assume f separates its critical points $z_1, \ldots, z_\rho$.

Put $f(z_i) = c_i$, and order the z_i so that $a < c_1 < \cdots < c_\rho < b$. Let $k(i)$ be the index of z_i. Choose regular values $a_0, \ldots, a_\rho$ so that

$$a = a_0; \qquad a_{i-1} < c_i < a_i; \qquad a_\rho = b.$$

Put $A = f^{-1}(a_0)$, $X_i = f^{-1}[a_0,a_i]$. Thus X_i is obtained from X_{i-1} by attaching a $k(i)$-cell. We denote this by

$$X_i = X_{i-1} \cup e^{k(i)}.$$

Define

$$\alpha(i,j) = \dim H_i(X_j, X_{j-1})$$

where homology groups always have coefficients in the field F. The excision axiom for homology implies

$$H_i(X_j, X_{j-1}) \approx H_i(D^{k(j)}, \partial D^{k(j)}).$$

Therefore

$$\alpha(i,j) = \begin{cases} 1 & \text{if} \quad i = k(j) \\ 0 & \text{if} \quad i \neq k(j). \end{cases} \tag{1}$$

Define

$$\beta(i,j) = \dim H_i(X_j, A).$$

Since X_j is obtained from A by adding cells of dimensions $\leqslant n$, $\beta(i,j) = 0$ for $i > n$.

Consider the exact homology sequence of the triple (X_j, X_{j-1}, A):

$$0 \to H_n(X_{j-1},A) \to H_n(X_j,A) \to H_n(X_j,X_{j-1}) \to \\ H_{n-1}(X_{j-1},A) \cdots \to H_0(X_j,X_{j-1}) \to 0.$$

Exactness implies the vanishing of the corresponding alternating sum of dimensions of the vector spaces in the sequence. Grouping 3 terms at a time in this sum gives:

$$0 = \sum_{i=0}^{n} (-1)^i[\beta(i,j-1) - \beta(i,j) + \alpha(i,j)].$$

Summing over j yields

$$\sum_{i=0}^{n} (-1)^i \left[\sum_{j=1}^{n} \alpha(i,j)\right] = \sum_{i=0}^{n} (-1)^i[\beta(i,n) - \beta(i,0)] \\ = \sum_{i=0}^{n} (-1)^i \beta_i$$

because $\beta(i,n) = \beta_i$ and $\beta(i,0) = 0$. Now

$$\sum_{j=1}^{n} \alpha(i,j)$$

is the number of $j \in \{1, \ldots, n\}$ such that $k(j) = i$, which is ν_i. Therefore

$$\sum_{i=0}^{n} (-1)^i \nu_i = \sum_{i=0}^{n} (-1)^i \beta_i.$$

This proves (b) of Theorem 3.3.

The proof of (a) is similar, starting from the exact sequence

$$0 \to K_{m,j} \to H_m(X_{j-1},A) \to H_m(X_j,A) \to H_m(X_j,X_{j-1})$$

where the first term is the kernel of $H_n(X_{j-1},A) \to H_n(X_j,A)$ and the rest of the sequence is as before. Let $\kappa_{m,j} = \dim K_{m,j}$. Exactness yields

$$\kappa_{m,j} = \sum_{i=0}^{m} (-1)^{i+m}[\beta(i,j-1) - \beta(i,j) + \alpha(i,j)].$$

Summing over $j = 1, \ldots, n$ gives

$$\sum_{j=1}^{n} \kappa_{m,j} + \sum_{i=0}^{m} (-1)^{i+m}\beta_i = \sum_{i=0}^{m} (-1)^{i+m} \sum_{j=1}^{n} \alpha(i,j) = \sum_{i=0}^{m} (-1)^{i+m}\nu_i$$

which implies (a) since $\kappa_{m,j} \geqslant 0$. The proof of Theorem 3.4 is complete.

QED

For the following important application of Theorem 3.5 let M be a compact manifold without boundary. Recall that in Chapter 4 we defined the Euler characteristic of M to be the sum of the indices of zeros of a vector field on M. We have also defined the *homological* Euler characteristic of M to be the alternating sum of the Betti numbers of M. The celebrated *Theorem of Hopf* equates these two characteristics:

3.8. Theorem. *The homological Euler characteristic equals the Euler characteristic for a compact manifold without boundary.*

Proof. The Euler characteristic can be computed from any vector field X on M having finitely many zeros. We choose $X = \frac{1}{2}\operatorname{grad} f$, where $f: M \to \mathbb{R}$ is a Morse function, and the gradient is taken in a Riemannian metric on M which, near each critical point, is induced from $\mathbb{R}^n$ by Morse coordinates.

Let p be a critical point of index k. Let $(x_1, \ldots, x_n)$ be Morse coordinates at $p = (0, \ldots, 0)$. Then

$$f(x) = -x_1^2 - \cdots - x_k^2 + x_{k+1}^2 + \cdots + x_n^2 + f(p),$$

and

$$X(x) = (-x_1, \ldots, -x_k, x_{k+1}, \ldots, x_n).$$

Thus X is the Cartesian product of the vector fields Y on $\mathbb{R}^k$, Z on $\mathbb{R}^{n-k}$ defined by $Y(y) = -y$, $Z(z) = z$. An easy computation shows that

$$\operatorname{Ind}_0 X = (\operatorname{Ind}_0 Y)(\operatorname{Ind}_0 Z)$$

and

$$\operatorname{Ind}_0 Y = (-1)^k, \qquad \operatorname{Ind}_0 Z = 1.$$

Therefore

$$\operatorname{Ind}_p X = (-1)^k.$$

Therefore the sum of the indices of zeros of X is $\sum_{k=0}^{n} (-1)^k v_k$ where v_k is the number of critical points of f of index k. By Theorem 3.5, this is the homological Euler characteristic.

QED

Exercises

1. The *ω-limit* set $L_\omega(x)$ of a point $x \in M$ for a vector field X on M is the set of points of the form $\lim_{n\to\infty} \eta(t_n, x)$ where η is the flow of X and $t_n \to \infty$. The *α-limit* set $L_\alpha(x)$ is defined similarly by letting $t_n \to -\infty$ instead of ∞.

(a) If M is compact, $L_\omega(x)$ and $L_\alpha(x)$ are connected, compact nonempty sets, invariant under the flow.

(b) If $X = \operatorname{grad} f$ and $f: M \to \mathbb{R}$ has isolated critical points then $L_\omega(x)$ and $L_\alpha(x)$ each consists of a single critical point.

2. Let $f: M \to \mathbb{R}$ be a C^{r+1} Morse function on a compact manifold. Suppose M has a C^∞ Riemannian metric which is induced by Morse charts near critical points. Consider the flow of grad f for this metric. If z is a critical point the *stable manifold* of z (also called the *inset*) is

$$W_s(z) = \{x \in M : L_\omega(x) = z\}$$

while the *unstable manifold* of z (also called the *outset*) is

$$W_u(z) = \{x \in M : L_z(x) = z\}.$$

(a) $W_s(z)$ and $W_u(z)$ are connected C^r submanifolds, of dimensions k and $n - k$ respectively, where $k = \text{Ind}(z)$.

(b) $W_s(z)$ and $W_u(z)$ intersect transversely at z and are otherwise disjoint.

*(c) If $\partial M = \varnothing$ then $W_s(z) \approx \mathbb{R}^{n-k}$ and $W_u(z) \approx \mathbb{R}^k$.

**(d) Actually, (a), (b) and (c) are true for any C^∞ Riemannian metric on M.

3. A compact n-manifold is the union of the closures of a finite number of disjoint open n-cells. [Consider stable manifolds of local minima of a Morse function.]

4. Let $f: M \to [a,b]$ be an admissible Morse function on a compact manifold, having a unique critical value c, $a < c < b$. Let $z_1, \ldots, z_m$ be the critical points of f and put $W_s = \bigcup_i W_s(z_i)$, $W_u = \bigcup_i W_u(z_i)$. (See Exercise 1 for definition of W_s and W_u.) Then:

$$M - (W_s \cup W_u) \approx [f^{-1}(a) - W_s] \times I \approx [f^{-1}(b) - W_u] \times I.$$

***5.** Let S be a compact surface of genus p.

(a) Every Morse function on S has at least $2p + 2$ critical points.

(b) Some Morse function has exactly this number of critical points.

6. Real projective n-space P^n can be represented as the quotient space of $\mathbb{R}^{n+1} - 0$ under the identifications $(x_0, \ldots, x_n) = (cx_0, \ldots, cx_n)$ if $c \neq 0$. The equivalence class of $(x_0, \ldots, x_n)$ is denoted by $[x_0, \ldots, x_n]$. Let $\lambda_0, \ldots, \lambda_n$ be distinct nonzero real numbers and define $f: P^n \to \mathbb{R}$ by

$$[x_0, \ldots, x_n] \mapsto \textstyle\sum_j \lambda_j x_j^2 / \sum_j x_j^2.$$

(a) f is a Morse function of type $(1,1, \ldots, 1)$. (See also Exercise 1, Section 6.4.)

(b) Sketch the critical points and level surfaces of the composition $S^n \xrightarrow{p} P^n \xrightarrow{f} \mathbb{R}$, where p is the canonical double covering, for $n = 1, 2, 3$.

***7.** Let $f: S^n \to \mathbb{R}$ be a Morse function invariant under the antipodal map $x \to -x$. Then f has at least one critical point of each index $0, 1, \ldots, n$. [Consider the function induced on P^n. The $\mathbb{Z}_2$ Betti numbers of P^n are $1, 1, \ldots, 1$.]

8. Let M be a compact n-manifold without boundary and $f: M \mapsto \mathbb{R}$ a C^2 map. Suppose every critical point belongs to a nondegenerate critical submanifold (see Exercise 7, Section 6.1). Let μ_k denote the sum of the Euler characteristics of the nondegenerate critical submanifolds of index k. Then the following generalized Morse equality holds:

$$\chi(M) = \sum_{k=0}^{n} (-1)^k \mu_k.$$

[Let $g: M \to \mathbb{R}$ vanish outside tubular neighborhoods of the critical submanifolds, such that g restricts to a Morse function on the critical submanifolds. One can choose g so that $f + g$ is a Morse function whose critical points are precisely those of g on the critical submanifolds. If V is a critical submanifold of index k and $x \in V$ is a critical point for g of index i, then x has index $(-1)^{k_i}$ as a critical point of $f + g$. The sum of the indices of the critical points of $f + g$ in V is thus $(-1)^k \chi(V)$. This result is due to R. Bott.]

9. Let $\varphi:M \to N$ be a surjective submersion where M and N are compact manifolds without boundary. Let $V = \varphi^{-1}(y)$ for some $y \in N$. Then $\chi(M) = \chi(V)\chi(N)$. [Let $g:N \to \mathbb{R}$ be a Morse function and apply Exercise 8 to $f = \varphi g:M \to \mathbb{R}$.]

10. Let $\partial_+ M$ be a union of components of the boundary of a compact n-manifold M; put $\partial_- M = \partial M - \partial_+ M$. Then

$$\chi'(M,\partial_+ M) = \chi'(M,\partial_- M).$$

4. *CW*-Complexes

In this section we assume familiarity with the notion of *CW-complex*. Briefly, a *CW*-complex is a space X which can be expressed $X = \bigcup_{n=0}^{\infty} X_n$ where $X_0 \subset X_1 \subset \ldots$, X_0 is a discrete subset and X_{n+1} is obtained from X_n by attaching n-cells by continuous maps of their boundaries into X_{n-1}. It is required that a subset of X is closed provided its intersection with each such cell (image) is closed. If $X = X_n$ we call X an n-dimensional *CW*-complex. If the number of cells is finite, X is a *finite* *CW*-Complex.

A *subcomplex* A of a *CW*-complex X is a closed subspace A which is the union of cells of X. A *CW-pair* (X,A) consists of a *CW* complex X and a subcomplex A.

4.1. Theorem. *Let M be a compact n-manifold and $f:M \to [a,b]$ an admissible Morse function of type $(\nu_0, \ldots, \nu_n)$ such that $\partial M = f^{-1}(b)$. Then M has the homotopy type of a finite CW complex having exactly ν_k cells of each dimension $k = 0, \ldots, n$ and no other cells.*

Proof. The proof is by induction on the number of critical values. If c_1 is the smallest critical value, then c_1 is the absolute minimum of f because $f^{-1}(a)$ is empty. Choose $a < c_1 < a_1$ so that c_1 is the only critical value in $[a,a_1]$. Then $f^{-1}[a,a_1]$ has the homotopy type of a finite discrete set of points by Theorem 3.4; in fact $f^{-1}[a_1 a_1]$ is the union of disjoint n-disks. This starts the induction. The inductive step follows from Theorem 3.4.

QED

Using the same ideas, one can prove the following result. The details are left to the reader.

4.2. Theorem. *Let M be a compact n-dimensional manifold. Then $(M,\partial M)$ has the homotopy type of a CW-pair of dimensions $\leqslant n$.*

The restriction that M be compact is not necessary, but the proof for noncompact manifolds uses triangulation. We outline a proof of a somewhat weaker result:

4.3. Theorem. *An n-dimensional manifold has the homotopy type of a CW-complex of dimension $\leqslant n$.*

Proof. One can show that M and $M - \partial M$ are homotopy equivalent (for example, by the collaring Theorem 2.1). Hence we can assume $\partial M = \varnothing$.

Choose a proper Morse function $f: M \to \mathbb{R}_+$, for example an approximation to the square of a proper map $M \to \mathbb{R}$. Assume f separates the critical points $z_1, z_2, \ldots$, which are ordered so that $f(z_{i+1}) > f(z_i)$. Choose numbers a_i so that

$$0 = a_0 < f(z_1) < a_1 < f(z_2) < \cdots.$$

Note that $f(z_i) \to \infty$ since f is proper and the z_i are isolated; therefore $a_i \to \infty$ also.

From Theorem 4.2, for each integer $k \geqslant 1$ there is a CW-complex X_k of dimension n and disjoint subcomplexes $Y_k, Z_k \subset X_k$ and a homotopy equivalence

$$u_k: f^{-1}[a_{k-1}, a_k] \to X_k,$$

taking $f^{-1}(a_{k-1})$ to Y_k and $f^{-1}(a_k)$ to Z_k by homotopy equivalences. Let $v_k: Z_k \to f^{-1}(a_k)$ be a homotopy inverse to $f: f^{-1}(a_k) \to Z_k$. The composition

$$Z_k \xrightarrow{v_k} f^{-1}(a_k) \xrightarrow{u_{k+1}} Y_{k+1}$$

can be approximated by a cellular map $w_k: Z_k \to Y_{k+1}$. A CW-complex X homotopy equivalent to M is obtained from the disjoint union of the X_k under the identification of $x \in Z_k$ with $w_k(x) \in X_{k+1}$.

QED

Another extension of Theorem 4.2 is:

4.4. Theorem. *Let M be a compact manifold and $A \subset M$ a compact submanifold, $\partial A = \partial M = \varnothing$. Then (M,A) has the homotopy type of a CW-pair.*

Proof. Let $N \subset M$ be a closed tubular neighborhood of A. Thus N is a closed submanifold with boundary admitting A as a deformation retract, and (M,A) has the same homotopy type as (M,N). Let $P = \overline{M - N}$; then $\partial P = \partial N = P \cup N$ and $P \cup N = M$. By Theorem 4.2, $(P,\partial P)$ and $(N,\partial N)$ have the homotopy type of CW-complexes; this implies Theorem 4.4.

QED

Again compactness is an unnecessary restriction. There are also generalizations of Theorem 4.4, to ∂-manifolds; for example A can be a closed neat submanifold.

Exercises

***1.** Let CP^n denote complex projective n-space. Define $g: CP^n \to \mathbb{R}$ by the formula

$$g[z_0, \ldots, z_n] = \sum \lambda_j |z_j|^2 / \sum |z_j|^2,$$

where $z_0, \ldots, z_n$ are complex homogeneous coordinates on CP^n and the λ_j are distinct positive numbers. Then g is a Morse function of type $(1,0,1,0, \ldots, 1,0,1)$. Therefore CP^n has the homotopy type of a CW complex with 1 cell in even dimensions $0, 2, \ldots, 2n$

and no other cells. Consequently

$$H_k(CP^n; G) = \begin{cases} G & \text{if} \quad 0 \leqslant k \leqslant 2n \text{ and } k \text{ is even} \\ 0 & \text{otherwise} \end{cases}$$

for any coefficient group G.

2. A *relative CW-complex* (X,A) consists of a space X and a closed subspace A such that $X = \bigcup_{n=-1}^{\infty} \bar{X}_n$, with $A = X_{-1} \subset X_0 \subset \ldots$, such that X_n is obtained from X_{n-1} by attaching n-cells. If $f:M \to [a,b]$ is an admissible Morse function of type $(v_0, \ldots, v_n)$ on a compact manifold M, then $(M, f^{-1}(a))$ has the homotopy type of a relative CW-complex having exactly v_k cells of dimension k, for each $k = 0, \ldots, n = \dim M$, and no other cells.

Chapter 7

Cobordism

. . . the theory of "Cobordisme" which has, within the few years of its existence, led to the most penetrating insights into the topology of differentiable manifolds.

—H. Hopf, International Congress
of Mathematicians, 1958

Mathematicians are like Frenchmen: whatever you say to them they translate into their own language and forthwith it is something completely different.

—Goethe, *Maximen und Reflexionen*

In this short chapter we present the elementary part of one of the most elegant theories in differential topology, René Thom's theory of cobordism. It was largely for this work that Thom was awarded the Fields Medal in 1958.

We can partition all compact n-manifolds without boundary into equivalence classes, two manifolds being equivalent if their disjoint union is the boundary of a compact $(n + 1)$-manifold. The set $\mathfrak{N}^n$ of equivalence classes becomes an abelian group under the operation of disjoint union. An analogous construction with oriented manifolds produces an abelian group Ω^n. Thom [1] set himself (and largely solved) the problem of computing these *cobordism groups*. Although their definition is very simple, it is not at all obvious how to compute them, or even to determine their cardinality.

Thom's work falls into two parts: first, proving that the cobordism groups are isomorphic to certain homotopy groups; and second, computing these homotopy groups. The second step requires a good deal of algebraic topology and we cannot go into it; it is the first step we are concerned with. Even though the proof yields no explicit calculations, it provides new insights into the connections between manifolds, vector bundles, homotopy and transversality.

1. Cobordism and Transversality

Two compact manifolds M_0, M_1 are called *cobordant*, denoted by $M_0 \sim M_1$, if there is a compact manifold W such that $\partial W \approx M_0 \times 0 \cup M_1 \times 1$. Speaking loosely, this means that the disjoint union of M_0 and M_1 is the boundary of W. We call W a cobordism from M_0 to M_1. It is easy to see that for each dimension n this defines an equivalence relation, called *cobordism*, on the class of compact n-manifolds without boundary. The set of cobordism classes is denoted by $\mathfrak{N}^n$; the cobordism class of M is $[M]$.

An analogous equivalence relation, *oriented cobordism*, is defined for oriented manifolds. Let ω_i be an orientation of M_i; then (M_0,ω_0) and (M_1,ω_1) are *cobordant* if there is a compact oriented manifold (W,θ) and an orientation preserving diffeomorphism

$$(\partial W,\partial\theta) \approx (M_0 \times 0, -\omega_0) \cup (M_1 \times 1, \omega_1).$$

The set of these equivalence classes is denoted by Ω^n.

1.1. Theorem. *The operation of disjoint union makes $\mathfrak{N}^n$ and Ω^n into abelian groups.*

Proof. First of all, diffeomorphic manifolds are cobordant: if $M_0 \approx M_1$ then

$$M_0 \times 0 \cup M_1 \times 1 \approx \partial(M_0 \times I)$$

(taking orientations into account where appropriate). The associative and commutative laws follow easily. The zero element of the group is $[V]$ for any V which is the boundary of some compact manifold W. For, taking M, $M \times I$ and W disjoint, we have

$$(M \times 0 \cup V) \cup (M \times 1) = \partial(M \times I \cup W).$$

Thus $M \cup V \sim M$. (We could take $V = \varnothing$.) The inverse of $[M]$ is $[M]$; the inverse of $[M,\omega]$ is $[M,-\omega]$, as is seen by looking at $M \times I$.

QED

We have defined two sequences of abelian groups, the *oriented* and *non-oriented cobordism groups*. How can they be computed? Since we know all manifolds of dimension 0 or 1, the computation is easy in this case (Exercise 3); and it is also simple in dimension 2 once we have classified surfaces. But this kind of head-on attack will not get very far.

For any topological problem it is always a good idea to try to bring maps into play. Now we have already seen a useful connection between maps and manifolds: if $f: V \to N$ is transverse to a submanifold $A \subset N$ then $f^{-1}(A)$ is a submanifold of V. If we fix V, N and A, and let f vary, we obtain a collection of submanifolds of V.

Once we consider maps, we should think about homotopy. A natural question is: if f, $g: V \to N$ are homotopic, how are $f^{-1}(A)$ and $g^{-1}(A)$ related? The following simple result is the key to cobordism:

1.2. Lemma. *Let V, N be manifolds without boundary and $A \subset N$ a closed submanifold without boundary. Assume V is compact. If $f,g: M \to N$ are homotopic maps, both of which are transverse to A, then the manifolds $f^{-1}(A)$, $g^{-1}(A)$ are cobordant.*

Proof. There is a homotopy $H: V \times I \to N$ from f to g, which we can choose transverse to A. Then $H^{-1}(A)$ is a cobordism from $f^{-1}(A)$ to $g^{-1}(A)$.

QED

We thus obtain a map from the homotopy set $[V,N]$ to $\mathfrak{N}^n$, where $n = \dim V - \dim N + \dim A$. If we want to capture all of $\mathfrak{N}^n$ in this way, we must choose V to be a manifold in which all n-manifolds can be embedded. Thus we take $V = S^{n+k}$ with large $k = \dim N - \dim A$. We now have a map $[S^{n+k},N] \to \mathfrak{N}^n$.

Suppose $f: S^{n+k} \to N$ is transverse to A. What can we say about the submanifold $f^{-1}(A) = M \subset S^{n+k}$? A most important fact is that *the normal bundle of M in S^{n+k} is the pullback under f of the normal bundle of A in N.* For by the definition of transversality, Tf induces a vector bundle map of algebraic normal bundles:

$$\begin{array}{ccc} T_M S^{n+k}/TM & \longrightarrow & T_A N/TA \\ \downarrow & & \downarrow \\ M & \xrightarrow[f]{} & A \end{array}$$

It follows that the submanifold $A \subset N$ must have the property that the normal bundle of any $M^n \subset S^{n+k}$ can be pulled back from the normal bundle of A by some map $M \to A$. Fortunately we have already constructed a pair (N,A) with this property, namely $(E_{s,k},G_{s,k})$ for $s > n + k$. Here $G_{s,k}$ is the Grassmann manifold of k-planes in $\mathbb{R}^s$ and $E_{s,k}$ is the total space of the universal bundle $\gamma_{s,k} \to G_{s,k}$ (see Section 4.3). As usual the base space is identified with the zero section, so that $G_{s,k} \subset E_{s,k}$.

For ease of notation we put $E = E_{s,k}$, $G = G_{s,k}$. Given a compact n-dimensional submanifold $M \subset S^{n+k}$ without boundary, let $U \subset S^{n+k}$ be a tubular neighborhood of M. Then, by Theorem 3.3 of Chapter 4, there is a map of vector bundles

$$\begin{array}{ccc} U & \xrightarrow{g} & E \\ \downarrow & & \downarrow \\ M & \xrightarrow[f]{} & G \end{array}$$

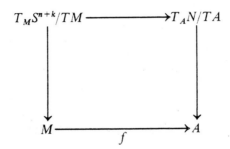

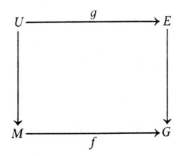

At this stage a new problem arises: we want g defined on all of S^{n+k}, not merely on the subset U. Moreover the extended map must not map any new points into G, since it is essential that $g^{-1}(G) = M$.

Simple examples show that such an extension may not be possible. At this point Thom introduces a *deus ex machina*: he adds a "point at infinity" to E and maps $S^{n+k} - U$ to the new point! This stroke of genius completely solves the problem; it only remains to work out the technical details. We do this in the next section.

Exercises

1. The operation of cartesian product of manifolds induces bilinear maps

$$\Omega^i \times \Omega^j \to \Omega^{i+j}.$$

These pairings are associative, and commutative in the sense of graded rings: if $\alpha \in \Omega^i$, $\beta \in \Omega^j$ the $\alpha\beta = (-1)^{i+j}\beta\alpha$. There are similar bilinear pairings

$$\mathfrak{N}^i \times \mathfrak{N}^j \to \mathfrak{N}^{i+j}$$

which are strictly commutative.

2. Every element of $\mathfrak{N}^i$ has order 2.

3. $\Omega^0 = \mathbb{Z}$, $\mathfrak{N}^0 = \mathbb{Z}_2$ and $\Omega^1 = \mathfrak{N}^1 = 0$.

4. Let $n \geqslant 0$ be an even integer. The "mod 2 Euler characteristic" defines a surjective homomorphism $\mathfrak{N}^n \to \mathbb{Z}_2$.

5. An orientable surface of genus $p \geqslant 0$ (as defined in Exercise 12, Section 1.3) is the boundary of a compact oriented 3-manifold.

2. The Thom Homomorphism

Let $\xi = (p,E,B)$ be a vector bundle over a compact manifold B without boundary. The *one-point compactification* E^* of E is the space $E^* = E \cup \{\infty\}$ where ∞ is a point not in E. Neighborhoods of ∞ are complements in E^* of compact subsets of E. We also call E^* the *Thom space* of the vector bundle ξ.

A fundamental property of E^* is that $E^* - B$ *is a contractible space.* A contraction to ∞ is given by the homotopy

$$(E^* - B) \times I \to E^* - B$$

$$(x,t) \mapsto \begin{cases} \left(\dfrac{1+t}{1-t}\right)x, & 0 \leqslant t < 1 \\ \infty, & t = 1 \end{cases}$$

From the homotopy extension theorem we obtain:

2.1. Lemma. *Let Y be a closed subset of a manifold Q. Then two maps $Q \to E^* - B$ which agree on Y are homotopic* rel Y.

Now let Q be a manifold and $h:Q \to E^*$ a map. We say that h is in *standard form* if there is a submanifold $M \subset Q$ and a tubular neighborhood $U \subset Q$ of M such that $U = g^{-1}(E)$, $M = g^{-1}(B)$, and the diagram

$$\begin{array}{ccc} U & \xrightarrow{g|U} & E \\ \downarrow & & \downarrow \\ M & \xrightarrow[g|M]{} & B \end{array}$$

is a vector bundle map. This implies: $g \pitchfork B$ and $g(Q - U) = \infty$.

2.2. Lemma. *Let Q be a compact manifold and B a compact manifold without boundary. Then every map $f:Q \to E^*$ is homotopic to a map in standard form.*

Proof. By a preliminary homotopy we assume $f \pitchfork B$. Put $M = f^{-1}(B)$; let $U \subset f^{-1}(E)$ be a tubular neighborhood of M and $D \subset U$ a disk subbundle. By Theorem 4.6.5 we can further assume that f agrees in D with a vector bundle map $\Phi: U \to E$. Define a map

$$h:Q \to E^*$$

$$h = \begin{cases} \Phi & \text{on} \quad U \\ \infty & \text{on} \quad Q - U. \end{cases}$$

Then h agrees with f on D, and h is in standard form. Since h and f agree on ∂D, and both map $Q - \operatorname{int} D$ into the contractible space $E^* - B$, it follows from Lemma 2.1 that $f \simeq h$.

QED

Let $E_{s,k} \to G_{s,k}$ be the Grassmann bundle; put $E = E_{s,k}$, $G = G_{s,k}$.

Let $\pi_{n+k}(E^*)$ denote the $(n + k)$'th homotopy group of $E^* = E^*_{s,k}$. The base point is unimportant; ∞ is the conventional choice. The homotopy class of $f:S^{n+k} \to E^*$ is denoted by $[f]$.

We now define the *Thom homomorphism* $\tau(n; k,s) = \tau:\pi_{n+k}(E^*) \to \mathfrak{R}^n$ as follows. Let $\alpha \in \pi_{n+k}(E^*)$. By the transversality theorem there exists $f \in \alpha$ such that $f:S^{n+k} \to E^*$ is transverse to G (or more precisely, $f|(S^{n+k} - f^{-1}(\infty))$ is transverse to G). By Theorem 2.1 the cobordism class of the manifold $f^{-1}(G)$ is independent of the choice of f. We define $\tau(\alpha) = [f^{-1}(G)]$. Thus $\tau(\alpha) = [f^{-1}(G)]$ where $f \in \alpha$, $f \pitchfork G$.

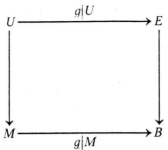

To see that τ is additive, let α, $\beta \in \pi_{n+1}(E^*)$. According to one of the definitions of addition in homotopy groups, we can define $\alpha + \beta$ as follows. Choose maps $f \in \alpha$, $g \in \beta$, such that f maps the lower hemisphere of S^{n+k} to ∞ and g maps the upper hemisphere to ∞. Then $\alpha + \beta$ is the class of the map h which equals f on the upper hemisphere and g on the lower hemisphere. We may assume that f, g and h are transverse to G. It is clear that $h^{-1}(G)$ is the disjoint union of $f^{-1}(G)$ and $g^{-1}(G)$, since these are in disjoint open hemispheres. Therefore

$$\begin{aligned}\tau(\alpha + \beta) &= [h^{-1}(G)]\\ &= [f^{-1}(G) \cup g^{-1}(G)]\\ &= [f^{-1}(G)] + [g^{-1}(G)]\\ &= \tau(\alpha) + \tau(\beta).\end{aligned}$$

For oriented manifolds there is defined a similar Thom homomorphism

$$\tilde{\tau}: \pi_{n+k}(\tilde{E}^*) \to \Omega^n.$$

Here $\tilde{E} = \tilde{E}_{s,k}$ is the total space of the vector bundle $\tilde{\gamma}_{s,k}$ defined as follows. The base space $\tilde{G} = \tilde{G}_{s,k}$ is the manifold of oriented k-planes in $\mathbb{R}^s$; an element of $\tilde{G}$ is a pair (P,ω) where $P \in G$ and ω is an orientation of the k-plane P. Thus $\tilde{G}$ is a double covering of G.

The vector bundle $\tilde{\gamma}_{s,k}$ is defined as the pullback of $\gamma_{s,k}$ by the covering map $\tilde{G} \to G$, $(P,\omega) \mapsto P$. Thus an element of $\tilde{E}$ is a triple (P,ω,x) comprising a k-plane $P \subset \mathbb{R}^s$, an orientation ω of P, and a vector $x \in P$. The bundle projection of $\tilde{\gamma}_{s,k}$ is the map

$$\tilde{E} \to \tilde{G}, \qquad (P,\omega,x) \mapsto (P,\omega).$$

Notice that the bundle $\tilde{\gamma}_{s,k}$ has a canonical orientation: the fibre over (P,ω) is oriented by ω.

The oriented Thom homomorphism $\tilde{\tau}$ is defined as follows. Let $\alpha \in \pi_{n+k}(\tilde{E}^*)$. Then $\tilde{\tau}(\alpha) = [f^{-1}(\tilde{G}),\theta]$ where

$$f: S^{n+k} \to \tilde{E}^*, \qquad f \pitchfork G, \qquad f \in \alpha$$

and the orientation θ of the manifold $M = f^{-1}(\tilde{G}) \subset S^{n+k}$ is defined as follows. The normal bundle ν of M is the pullback of the normal bundle of $\tilde{G}$ in $\tilde{E}$, which is just $\tilde{\gamma}_{s,k}$. Orient ν by the pullback of the canonical orientation of $\tilde{\gamma}_{s,k}$. Give S^{n+k} its standard orientation σ; and then give M the unique orientation θ such that $\theta \oplus \nu = \sigma$ on $T_M S^{n+k}$.

The following is the *fundamental theorem of cobordism*:

2.3. Theorem (Thom). *The Thom homomorphisms $\tau(n; k,s)$ and $\tilde{\tau}(n; k,s)$ are*:

(a) *surjective if $k > n$ and $s \geqslant k + n$;*
(b) *injective (and hence bijective) if $k > n + 1$ and $s \geqslant k + n + 1$.*

Proof. For simplicity we deal only with the unoriented case. To prove (a) let $[M^n] \in \mathfrak{N}^n$. We can assume $M^n \subset S^{n+k}$ by Whitney's embedding theorem. Let $U \subset S^{n+k}$ be a tubular neighborhood of M^n. By IV. 3.4 there is a vector bundle map $U \to E_{s,k}$, because $s \geqslant k + n$. Extend this to the map $g: S^{n+k} \to E^*_{s,k}$ which sends $S^{n+k} - U$ to ∞. Clearly $g \pitchfork G_{s,k}$ and $g^{-1}(G_{s,k}) = M^n$. Hence $\tau([g]) = [M^n]$.

To prove (b), suppose $g: S^{n+k} \to E^*_{s,k}$ is such that $\tau([g]) = 0 \in \mathfrak{N}^n$. We may assume g is in standard form by Theorem 2.2. Put $g^{-1}(G_{s,k}) = M^n$. Then $\tau([g]) = [M^n] = 0$, so that M bounds a compact manifold W^{n+1}. The assumption $k > n + 1$ implies that the inclusion $M^n \to S^{n+k}$ extends to a neat embedding $W^{n+1} \to D^{n+k+1}$.

The tubular neighborhood $U \subset S^{n+k}$ of M^n (used for the standard form of g) extends to a tubular neighborhood $V \subset D^{n+k+1}$ of W^{n+1} (see Theorem 4.6.4). By Theorem 4.3.4 the bundle map $g: U \to E_{s,k}$ extends to a bundle map $h: V \to E_{s,k}$, since $s \geqslant k + (n + 1)$. We extend h over all of D^{n+k+1} by mapping $D^{n+k+1} - V$ to ∞. Since $h|S^{n+k} = g$ it follows that $[g] = 0$.

QED

As was mentioned earlier we cannot go into the actual computation of cobordism groups, nor can we discuss their important applications; but the following remarks may indicate some of the power of Thom's theory.

It is not hard to show that E^* and $\tilde{E}^*$ are finite simplicial complexes. It follows from the simplicial approximation theorem that they have countably generated homotopy groups. Therefore the groups $\mathfrak{N}^n$ and Ω^n are countably generated—a conclusion by no means obvious from their definitions. Using algebraic topology it can be shown that in fact these groups are *finitely* generated. This means that there is a finite set $\mathscr{S}$ of n-manifolds having the following property: every compact n-manifold without boundary is cobordant to the disjoint union of a finite number of copies of elements of $\mathscr{S}$.

Much sharper results on the nature of the cobordism groups are known. As a sample we quote without proof a truly remarkable theorem of Thom:

2.4. Theorem. *Let n be a positive integer.*

(a) *If n is not divisible by* 4 *then the oriented cobordism group Ω^n is finite.*

(b) *If n is divisible by* 4, *say $n = 4k$, then Ω^n is a finitely generated abelian group whose rank is $\pi(k)$, the number of partitions of k. Moreover:*

(c) *A basis for the torsion-free part of Ω^{4k} consists of all products of the form $CP^{2j_1} \times \cdots \times CP^{2j}$ with $j_1 + \cdots + j_r = k$ and $1 \leqslant j_1 \leqslant \cdots \leqslant j_r \leqslant k$.*

Thus the "differentiable" problem of computing cobordism groups has been largely reduced to the combinatorial problem of computing $\pi(k)$. This latter problem is classical; unfortunately it has not been solved.

Exercises

1. (a) The space $E^*_{n,1}$ is homeomorphic to P^n.

(b) The space $\tilde{E}^*_{n,1}$ is homeomorphic to the space obtained by identifying two points of S^n.

(c) The space $E^*_{n,n}$ is homeomorphic to S^n.

2. Let E be the total space of a k-plane bundle over a compact manifold.

(a) If M is a manifold of dimension less than k, every map $M \to E^*$ is homotopic to a constant.

(b) The inclusion $\mathbb{R}^k \to E$ of a fibre extends to a map $S^k \to E^*$ which is not homotopic to a constant.

3. Two n-dimensional submanifolds M_0, M_1 of a compact manifold V are called *V-cobordant* if there is a compact submanifold $W \subset V \times I$ such that

$$\partial W = M_0 \times 0 \cup M_1 \times 1.$$

This is an equivalence relation; the set of equivalence classes is denoted $\mathfrak{N}^n(V)$.

*(a) Let $\dim V = n + k$. The natural map $\tau_v:[V,E^*_{s,k}] \to \mathfrak{N}^n(V)$ is surjective if $k > n$, $s \geqslant k + n$ and bijective if $k > n + 1$, $s \geqslant k + n + 1$.

(b) Under what assumptions does the operation "disjoint union" make $\mathfrak{N}^n(V)$ a semigroup? A group?

(c) $\mathfrak{N}^n(S^{n+k})$ is a group for all n, k.

4. Let V be a Riemannian manifold. A *framed submanifold* (M^n,F) of V^{n+k} is a compact submanifold $M^n \subset V$, plus a family $F = (F_1, \ldots, F_k)$ of sections of $\tau_M V$ such that $(F_1(x), \ldots, F_k(x))$ are independent vectors spanning a subspace transverse to M_x, for all $x \in M$. Two framed submanifolds (M_0,F), (M_1,F') are *framed cobordant* if there is a framed submanifold $(W,G) \subset V \times I$ such that $\partial W = M_0 \times 0 \cup M_1 \times 1$, and $G|\partial W = F \cup F'$. The resulting set of equivalence classes is denoted $F^n(V^{n+k})$.

(a) There is a natural map

$$\pi:[V^{n+k},S^k] \to F^n(V^{n+k}).$$

[Imagine $S^k = (\mathbb{R}^k)^* = (E_{k,k})^*$.]

(b) π is an isomorphism for all k, n.

(c) When $k = 0$ and V^n is connected and oriented, there is a "degree" isomorphism $F^n(V^n) \approx \mathbb{Z}$. Thus we recapture the isomorphism $\deg:[V^n,S^n] \approx \mathbb{Z}$.

5. Let $\eta = (p,E,B)$ be a fixed vector bundle over a compact manifold B, $\partial B = \varnothing$. An *η-submanifold* $(M,f) \subset V$ of a manifold V is a pair (M,f) where $M \subset V$ is a compact submanifold and f is a bundle map from the normal bundle of M to η (this requires $\dim \eta = \dim V - \dim M$). Two η-submanifolds $(M_i,f_i) \subset V$ are *η-cobordant* if there is an η-submanifold $(W,f) \subset V \times I$ such that $\partial(W,f) = (M_0,f_0) \times 0 \cup (M_1,f_1) \times 1$ (using an obvious notation). The set of η-cobordism classes corresponds bijectively to the homotopy set $[V,E^*]$.

6. The *bordism group* $\Omega^n(X)$ of a space X is defined as follows. An element of $\Omega^n(X)$ is an equivalence class $[f,M]$ of maps $f:M \to X$ where M is a compact oriented n-manifold without boundary. Two maps equivalent if they extend to a map defined on an oriented cobordism between their domains. Taking X = a point gives Ω^n. A homotopy class of maps $g:X \to Y$ induces a homomorphism of abelian groups $g_\#:\Omega^n(X) \to \Omega^n(Y)$, by composition with maps $f:M \to X$.

***7.** There are natural homomorphisms $\Omega^n(X) \to H_n(X)$. For $n = 1$ these are isomorphisms.

8. There is a bilinear pairing

$$\Omega^n(X) \times \Omega^m(X) \to \Omega^{n+m-p}(X)$$

induced by intersection of maps, when X is an oriented p-dimensional manifold.

Chapter 8

Isotopy

Let us think, say, of a surface or a solid made of rubber, with figures marked upon it. What is preserved in these figures if the rubber is arbitrarily distorted without being torn?

—F. Klein, *Elementary Mathematics from an Advanced Standpoint*, 1908

In this chapter we investigate more thoroughly the notion of isotopy, introduced earlier for the tubular neighborhood theorem.

Intuitively speaking, we call two embeddings $f, g: V \hookrightarrow M$ isotopic if one can be deformed to the other through embeddings; such a deformation is called an isotopy. By itself this relation is not very useful. However, it is usually true that the isotopy can be realized by a diffeotopy of M, that is, by a one-parameter family h_t of diffeomorphisms of M such that $h_0 = 1_M$ and $h_1 f = g$. In this case f and g embed V in M "in the same way". It follows, for example, that if f extends to an embedding $F: W \hookrightarrow M$, where $W \supset V$, then g also extends to an embedding of W, namely $h_1 F$.

Thus in order to extend an embedding, it suffices to prove it is isotopic to an extendable embedding. This extension technique is one of the main uses of isotopy.

In Section 8.1 we prove the fundamental isotopy extension Theorem 1.3, along with several variations and applications. Section 8.2 applies these results to the question of differential structure on the union of two smooth manifolds which have been glued together along boundary components. In Section 8.3 special isotopies are constructed for embeddings of disks, the point being that there is only one way, up to isotopy and orientation, of embedding a disk in a connected manifold. Diffeotopies of the circle are also treated.

These results will be used in Chapter 9 to classify compact surfaces. They are a basic tool in any attempt to analyze manifolds or embeddings. Several applications are given in the exercises.

1. Isotopy

Let V and M be manifolds. Recall that an *isotopy* from V to M is a map $F: V \times I \to M$ such that for each $t \in I$ the map

$$F_t: V \to M, \qquad x \mapsto F(x,t)$$

is an embedding. Intuitively, an isotopy is a smooth 1-parameter family of embeddings.

The *track* of the isotopy F is the embedding

$$\hat{F}: V \times I \to M \times I,$$
$$(x,t) \mapsto (F(x,t),t).$$

Notice that F is *level-preserving*—it preserves the coordinate t. Every level-preserving embedding is the track of an isotopy.

If $F: V \times I \to M$ is an isotopy we call the two embeddings F_0 and F *isotopic*: we also say that F *is an isotopy of* F_0. If V is a submanifold of M and F_0 is the inclusion, we call F an *isotopy of* V *in* M. When $V = M$ and each F_t is a diffeomorphism, and $F_0 = 1_M$, then F is called a *diffeotopy* or an *ambient isotopy*.

There is an important connection between diffeotopies of M and vector fields on $M \times I$. Let $\hat{F}: M \times I \to M \times I$ be the track of a diffeotopy F, so that $\hat{F}$ is a level-preserving diffeomorphism. Each point of $M \times I$ belongs to a unique arc $\hat{F}(x \times I)$ for some $x \in M$. The tangent vectors to these arcs form a nonvanishing vector field X_F on $M \times I$, which is carried by the projection $M \times I \to I$ to the constant positive unit vector field on I. Thus there is a map $H: M \times I \to TM$ such that

$$X_F(y,t) = (H(y,t),1) \in M_y \times \mathbb{R} = T_{(y,t)}(M \times I).$$

The isotopy F is the flow Φ of X_F applies to $M \times 0$:

$$\begin{array}{ccc} M & = & M \times 0 \\ {\scriptstyle F_t}\downarrow & & \downarrow{\scriptstyle \Phi_t} \\ M & = & M \times t \end{array}$$

The *horizontal part* H of X_F is a special case of a *time-dependent vector field* on M. By this is meant any map $G: M \times I \to TM$ such that $G(x,t) \in M_x$; we also require that G map $\partial M \times I$ into $T(\partial M)$.

Not every time-dependent vector field G comes from a diffeotopy, for there is no guarantee that the flow of the corresponding vector field X on $M \times I$ is defined for all $t \in I$. The following diagram (Figure 8–1) shows the solution curves of a vector field on $\mathbb{R} \times I$ which can be scaled to have vertical component 1; but *no* solution curve goes from $\mathbb{R} \times 0$ to $\mathbb{R} \times 1$!

Since time-dependent vector fields are easy to construct, it is useful to have a criterion which guarantees that they generate isotopies. One such condition is the following. A time-dependent vector field $G: M \times I \to TM$ has *bounded velocity* if M has a complete Riemannian metric such that $|G(x,t)| < K$ for some constant K. (Compare Exercise 1 of Section 6.2.)

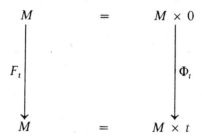

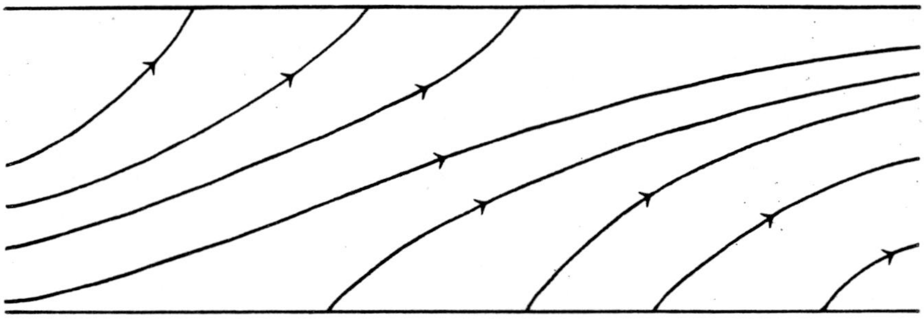

Figure 8–1. A vector field on $\mathbb{R} \times I$.

1.1. Theorem. *Let G be a time-dependent vector field on M having bounded velocity. Then G generates a diffeotopy of M. That is, there is a unique diffeotopy $F: M \times I \to M$ such that*

$$\frac{\partial F}{\partial t}(x,t) = G(F(x,t),t).$$

Proof. Let $X: M \times I \to T(M \times I)$ be the vector field $X(x,t) = (G(x,t),1)$. The projection into I of a solution curve of X is a curve of the form $y \mapsto y + t$. Therefore all solution curves are defined on intervals of length $\leqslant 1$. The condition of bounded velocity implies that M has a complete Riemannian metric in which all solution curves have finite length. Completeness then implies that each solution curve lies in a compact set. This means solution curves are defined on closed finite intervals, the endpoints of which map into $M \times 0$ and $M \times 1$. It follows that for $x \in M$ there is a solution curve of X having the form

$$t \mapsto (F(x,t),t), \qquad 0 \leqslant t \leqslant 1.$$

This defines the diffeotopy F. Uniqueness of F follows from uniqueness of solutions of Lipschitz differential equations.

QED

The *support* of a time-dependent vector field $G: M \times I \to TM$ is the set $\operatorname{Supp} G \subset M$ which is the closure of

$$\{x \in M: G(x,t) \neq 0 \qquad \text{for some} \qquad t \in I\}.$$

If $\operatorname{Supp} G$ is compact then G has bounded velocity. Therefore an immediate consequence of Theorem 1.1 is:

1.2. Theorem. *A time-dependent vector field which has compact support generates an isotopy. In particular every time-dependent vector field on a compact manifold generates an isotopy.*

The *support* $\operatorname{Supp} F \subset V$ of an isotopy $F: V \times I \to M$ is the closure of $\{x \in V: F(x,t) \neq F(x,0) \text{ for some } t \in I\}$.

We can now prove the following *isotopy extension theorems*:

1.3. Theorem. *Let $V \subset M$ be a compact submanifold and $F: V \times I \to M$ an isotopy of V. If either $F(V \times I) \subset \partial M$ or $F(V \times I) \subset M - \partial M$, then F extends to a diffeotopy of M having compact support.*

1.4. Theorem. *Let $U \subset M$ be an open set and $A \subset U$ a compact set. Let $F: U \times I \to M$ be an isotopy of U such that $\hat{F}(U \times I) \subset M \times I$ is open. Then there is a diffeotopy of M having compact support, which agrees with F on a neighborhood of $A \times I$.*

Proof of Theorems 1.3 and 1.4. We first prove Theorem 1.4. The tangent vectors to the curves

$$\hat{F}: x \times I \to M \times I \qquad (x \in U)$$

define a vector field X on $\hat{F}(U \times I)$ of the form $X(y,t) = (H(y,t),1)$. Here $H: \hat{F}(U \times I) \to TM$ with $H(y,t) \in M_y$. By means of a partition of unity we construct a time-dependent vector field $G: M \times I \to TM$ which agrees with H on a neighborhood of $A \times I$. (This requires $\hat{F}(U \times I)$ to be open.) Since $A \times I$ is compact, we can make G have compact support. The required diffeotopy of M is that generated by G.

To prove Theorem 1.3 we start from the vector field X on $\hat{F}(V \times I)$ tangent to the curves $\hat{F}(x \times I)$. By means of a tubular neighborhood of $\hat{F}(V \times I)$ and a partition of unity, the horizontal part of X is extended to a vector field Y on a neighborhood of $\hat{F}(V \times I)$ in $M \times I$. The hypothesis on F allows us to assume that $Y_{(x,t)}$ is tangent to $(\partial M) \times I$ whenever $x \in \partial M$. After restricting to a smaller neighborhood, the horizontal part of Y is extended to a compactly supported time-dependent vector field G on M. The diffeotopy generated by G completes the proof of Theorem 1.3.

QED

The following is a frequently used corollary of the isotopy extension theorem.

1.5. Theorem. *Let $V \subset N$ be a compact submanifold. Let $f_0, f_1: V \hookrightarrow M - \partial M$ be embeddings which are isotopic in $M - \partial M$. If f_0 extends to an embedding $N \to M$ then so does f_1.*

Proof. There is an isotopy from the inclusion $f_0(V) \subset M - \partial M$ to $f_1 f_0^{-1}: f_0(V) \hookrightarrow M - \partial M$. Such an isotopy extends to a diffeotopy H of M by 1.3. Thus $H_1: M \to M$ is a diffeomorphism such that $H_1 | f_0(V) = f_1^{-1} f_0$, or equivalently $H_1 f_0 = f_1$. Therefore if $g: N \hookrightarrow M$ extends f_0, then $H_1 g: N \hookrightarrow M$ is an embedding which extends f_1.

QED

Compactness in Theorems in 1.3 and 1.4 can be replaced with the weaker hypothesis of bounded velocity of the isotopy. If $V \subset M$ is a submanifold, an isotopy $F: V \times I \to M$ has *bounded velocity* if M has a complete metric with the property that the tangent vectors to the curves $t \mapsto F(x,t)$, have bounded lengths. We obtain:

1.6. Theorem. *Let $V \subset M$ be a closed submanifold and $F: V \times I \to M$ an isotopy of V having bounded velocity. If either $F(V \times I) \subset \partial M$ or $F(V \times I) \subset M - \partial M$, then F extends to a diffeotopy of M which has bounded velocity.*

1.7. Theorem. *Let $A \subset M$ be a closed set and $U \subset M$ an open neighborhood of A. Let $F: U \times I \to M$ be an isotopy of U having bounded velocity, such that $\hat{F}(U \times I)$ is open in $M \times I$. Then there is a diffeotopy G of M having bounded velocity, which agrees with F on a neighborhood of $A \times I$; and* $\text{Supp}\, G \subset F(U \times I)$.

The proofs are left to the reader.

As a corollary of Theorem 1.7 we obtain the *ambient tubular neighborhood theorem*:

1.8. Theorem. *Let $A \subset M$ be a closed neat submanifold and let $U \subset M$ be a neighborhood of A. Then the A-germ of any isotopy of tubular neighborhoods of A extends to a diffeotopy of M having support in U.*

Proof. Since an isotopy of tubular neighborhoods leaves A pointwise fixed, in some neighborhood of A it has bounded velocity and we can use Theorem 1.7.

QED

A theorem analogous to Theorem 1.8 holds for collars on ∂M. It has as a consequence the following *smoothing theorem*, which allows us to change certain kinds of homeomorphisms into diffeomorphisms (see Figure 8–2):

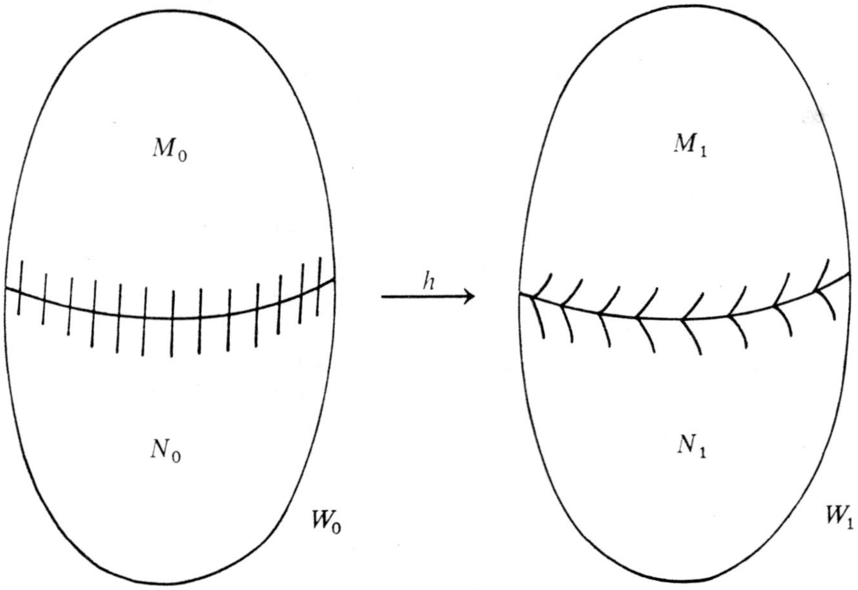

Figure 8–2. Smoothing the homeomorphism h.

1.9. Theorem. *For $i = 0, 1$ let W_i be an n-manifold without boundary which is the union of two closed n-dimensional submanifolds M_i, N_i such that*

$$M_i \cap N_i = \partial M_i = \partial N_i = V_i.$$

Let $h: W_0 \to W_1$ be a homeomorphism which maps M_0 and N_0 diffeomorphically onto M_1 and N_1 respectively. Then there is a diffeomorphism $f: W_0 \approx W_1$ such that $f(M_0) = M_1$, $f(N_0) = N_1$ and $f|V_0 = h|V_0$. Moreover f can be chosen so as to coincide with h outside a given neighborhood Q of V.

Proof. Choose a tubular neighborhood τ_i for V_i in W_i. This defines a collar on $\tau_i|M_i$ on V_i in M_i, and $\tau_i|N_i$ on V_i in N_i. We have another collar $h(\tau_0|M_0)$ on V_1 in M_1, which is the collar induced from $\tau_0|M_0$ by $h|M_0$. By the ambient tubular neighborhood Theorem 1.8 we can isotop $h|M_0: M_0 \to M_1$ to a new diffeomorphism $f': M_0 \to M_1$, $f' = h$ on V_0 and on $M_0 - Q$, such that $f'(\tau_0|M_0)$ has the same V_1-germ as $\tau_1|M_1$. Similarly we can isotop $h|N_0: N_0 \to N_1$ to $f'': N_0 \to N_1$ so that $f'' = h$ on V_0 and on $N_0 - Q$, and the collar $f''(\tau_0|N_0)$ has the same V_1-germ as $\tau_1|N_1$. The map $f' \cup f'': W_0 \to W_1$ is then the required diffeomorphism f.

QED

By choosing collars more carefully, we can even make $f = h$ on M_0 (or on M_1).

Exercises

1. The relation "f is isotopic to g" is an equivalence relation on $\mathrm{Emb}^\infty(M,N)$.

2. If $f_0, f_1: M \hookrightarrow N$ are isotopic and $g_0, g_1: N \hookrightarrow W$ are isotopic, then $g_0 f_0, g_1 f_1: M \subset W$ are isotopic.

3. (a) If $F: M \times I \to N$ is an isotopy, the map $I \to \mathrm{Emb}_W^\infty(M,N)$, $t \mapsto F_t$, is continuous.
(b) Conversely, every continuous map $\lambda: I \to \mathrm{Emb}_W^\infty(M,N)$ can be approximated by maps μ such that $\mu(i) = \lambda(i)$, $i = 0, 1$, and the map $M \times I \to N$, $(x,t) \mapsto \mu(t)(x)$ is an isotopy.
(c) Part (b), but not (a), is true for $\mathrm{Emb}_S^\infty(M,N)$.

4. The theorems of Section 8.1 are true for C^r isotopies (supply the definition) and C^r vector fields, $1 \leqslant r < \infty$. But some of them are false for C^ω isotopies.

*5. The equivalence classes of the relation "f is isotopic to g," on $\mathrm{Emb}^\infty(M,V)$, are open sets in the strong topology.

6. Let $V \subset M$ be a submanifold. A *k-isotopy* of V ($k = 2, 3, \ldots$) is a (C^∞) map $F: V \times I^k \to M$ such that for each $t \in I^k$ the map $F|V \times t$ is an embedding. Similarly one defines *k-diffeotopy* of M. If $V \subset M - \partial M$ and V is compact, then every k-isotopy of V extends to a k-diffeotopy of M with "compact support."

7. (a) Let M be a compact n-manifold without boundary and $f: M \hookrightarrow \mathbb{R}^{n+1}$ an embedding. If $L \in GL(n+1)$ has negative determinant then f is not isotopic to $L \circ f$. [Hint: Consider the degree of the *Gauss map* $\gamma: M \to S^n$ where $\gamma(x)$ is the outward unit normal vector at $x \in M$.]

(b) The inclusion of $S^n \hookrightarrow \mathbb{R}^{n+1}$ is isotopic to the antipodal embedding $(x \mapsto -x)$ if and only if n is odd.

8. Suppose ∂M^n is compact. If M^n embeds in $\mathbb{R}^q$, $q \geqslant 2n$ then every embedding $\partial M^n \to \mathbb{R}^q$ extends to M^n.

***9.** Let $L \subset \mathbb{R}^3$ be obtained from a straight line by putting a small knot in it thus:

(a) There is an isotopy F_t of L in $\mathbb{R}^3$ such that $F_1(L)$ is a straight line R. ["Roll" the knot to infinity.]

(b) Such an isotopy cannot be ambient because $\mathbb{R}^3 - L$ and $\mathbb{R}^3 - R$ have different fundamental groups.

10. Let $f, g: M \hookrightarrow V$ be homotopic embeddings. If $1 + \dim V > 2(1 + \dim M)$, then f and g are isotopic. [Let $f: M \times I \to V$ be a homotopy from f to g. Approximate the map $M \times I \to V \times I$, $(x,t) \mapsto (F(x,t),t)$ by an embedding H. Write $H(x,t) = (G(x,t), K(x,t)) \in V \times I$. Then, assuming $H(x,i) = F(x,i)$ for $i = 0, 1$, the map $G: M \times I \to V$ is an isotopy from f to g.]

***11.** Let M, V be noncompact manifolds, $\partial M = \partial V = \emptyset$. Let $f, g: M \subset V$ be embeddings that are homotopic by a proper map $M \times I \to V$. If $1 + \dim V > 2(1 + \dim M)$ then there is a proper ambient isotopy from f to g.

*****12.** Can the dimension restriction in Exercise 11 be weakened?

13. Let M be a compact submanifold of Q. Suppose $\partial Q = \emptyset$ and $\dim Q \geqslant 2 \dim M + 2$. If M is contractible to a point in Q then:

(a) M can be isotoped into any open subset of Q; and
(b) M lies in a coordinate domain.

14. Let $M, N \subset S^d$ be disjoint compact submanifolds. Suppose $\dim M + \dim N < d - 1$. Then M and N can be *geometrically separated* by an isotopy. This means that there is a diffeotopy of S^d carrying M into the northern hemisphere E^d_+ and N into the southern hemisphere E^d_-. [Assume $d \geqslant 2 \dim M + 2$. Use Exercise 13 to isotop M into E^d_+. By general position choose the isotopy to avoid N. Extend to an ambient isotopy of $S^d - N$ having compact support; etc.]

15. This exercise outlines a geometric proof of the "easy part" of the celebrated *Freudenthal suspension theorem* of homotopy theory: the suspension homomorphism $\Sigma: \pi_m(S^q) \to \pi_{m+1}(S^{q+1})$ is surjective if $m < 2q$ and injective if $m < 2q - 1$. Here $\pi_m(S^q)$ is the set of homotopy classes of maps $S^m \to S^q$ (the group structure is irrelevant). Σ is defined as follows.

Given $F: S^m \to S^q$, let $\Sigma f: S^{m+1} \to S^{q+1}$ coincide with f on the equator and map the north and south poles S^{m+1} to the corresponding poles of S^{q+1}; and let Σf map each great circle quadrant of S^{m+1}, joining a pole to the equator, isometrically onto a great circle quadrant of S^{q+1}.

(a) A map $g: S^{m+1} \to S^{q+1}$ is homotopic to a suspension if

$$g(E^{m+1}_+) \subset E^{q+1}_+ \quad \text{and} \quad g(E^{m+1}_-) \subset E^{q+1}_-.$$

(b) A map $g: S^{m+1} \to S^{q+1}$ is homotopic to a suspension if g^{-1}(north pole) $\subset$ int E^{m+1}_+ and g^{-1}(south pole) $\subset$ Int E^{m+1}_-. (For then $g(E^{m+1}_+) \subset S^{q+1} -$ (south pole) which deforms into E^{q+1}_+; etc.)

(c) If $m < 2q$ then $\sum:\pi_m(S^q) \to \pi_{m+1}(S^{q+1})$ is surjective. [Assume the poles of S^{q+1} are regular values; use (b) and Exercise 14.]

(d) If $m < 2q - 1$ then $\sum:\pi_m(S^q) \to \pi_{m+1}(S^{q+1})$ is injective. [Imitate the proof of (c) to show that a homotopy from $\sum(f)$ to $\sum(g)$ is homotopic to the suspension of a homotopy from f to g.]

16. The conclusion of Theorem 1.5 is true for isotopic embeddings $V \hookrightarrow M$.

***17.** Figure 1–5 shows three surfaces in $\mathbb{R}^3$. There are diffeotopies of $\mathbb{R}^3$ carrying any one of them onto any other!

2. Gluing Manifolds Together

Suppose that P and Q are n-dimension ∂-manifolds and that $f:\partial Q \approx \partial P$ is a diffeomorphism. The adjunction space $W = P \bigcup_f Q$ is a topological manifold containing natural copies of P and Q. We can give W a differential structure which extends the differential structures on P and Q. The object of this section is to show that all such differential structures on W are diffeomorphic.

For notational simplicity we identify P and Q with their images of W. Let $\partial P = \partial Q = V$. By means of collars on V in P and Q, we find a homeomorphism of a neighborhood $U \subset W$ of V onto $V \times \mathbb{R}$ taking $x \in V$ to $(x,0)$, and which maps $U \cap P$ and $U \cap Q$ diffeomorphically onto $V \times [0,\infty)$ and $V \times (-\infty,0]$, respectively. We give U the differential structure induced by this homeomorphism. The required differential structure on W is obtained by collation from P, Q, and U.

In defining this differential structure on W, various choices were made. The following result, called *uniqueness of gluing*, says that the diffeomorphism type of W is independent of these choices. It merely restates Theorem 1.9.

2.1. Theorem. *Let $f:\partial Q \approx \partial P$ be a diffeomorphism. Let α, β be two differential structures on $W = P \bigcup_f Q$ which both induce the original structure on P and Q. Then there is a diffeomorphism $h: W_\alpha \approx W_\beta$ such that $h|P = 1_P$.*

This theorem is somewhat unsatisfying in that there is no *canonical* differential structure on $P \bigcup_f Q$; there is only a canonical *diffeomorphism class* of structures. Differential topologists generally ignore this, and treat $P \bigcup_f Q$ as a well-defined differentiable manifold. Since it leads to no trouble and saves a good deal of writing, we shall follow this practice.

The following is a useful criterion for diffeomorphism of two glued manifolds:

2.2. Theorem. *Let $f_0:\partial Q_0 \approx \partial P$ and $f_1:\partial Q_1 \approx \partial P$ be diffeomorphisms. Suppose that the diffeomorphism $f_1^{-1}f_0:\partial Q_0 \approx \partial Q_1$ extends to a diffeomorphism $h:Q_0 \approx Q_1$. Then $P \bigcup_{f_0} Q_0 \approx P \bigcup_{f_1} Q$.*

Proof. A map

$$\psi:P \bigcup_{f_0} Q_0 \to P \bigcup_{f_1} Q_1$$

is well-defined by $\psi|P = 1_P$, $\psi|Q_0 = h$. Now apply Theorem 1.9 and the remark following it.

QED

An important special case is:

2.3. Theorem. *Let* $f, g: \partial Q \approx \partial P$ *be isotopic diffeomorphisms. Then* $P \bigcup_f Q \approx P \bigcup_g Q$.

Proof. $g^{-1}f$ is isotopic to the identity of ∂Q. The isotopy can be spread out over a collar on ∂Q and then extended to a diffeomorphism of Q which is the identity outside the collar. Now use Theorem 2.3.

QED

3. Isotopies of Disks

The following useful result says that, except perhaps for orientation, there is essentially only one way to embed a disk in a connected manifold.

3.1. Theorem. *Let* M *be a connected n-manifold and* $f, g: D^k \subset M$ *embeddings of the k-disk,* $0 \leqslant k \leqslant n$. *If* $k = n$ *and* M *is orientable, assume that* f *and* g *both preserve, or both reverse, orientation. Then* f *and* g *are isotopic. If* $f(D^k) \cup g(D^k) \subset M - \partial M$, *an isotopy between them can be realized by a diffeotopy of* M *having compact support.*

Proof. We shall use repeatedly the fact that isotopy is an equivalence relation on the set of embeddings.

First assume $\partial M = \varnothing$.

Since M is connected, the embeddings $f|0$, $g|0: 0 \to M$ are isotopic; by Theorem 1.3 they are ambiently isotopic. Therefore we may assume $f(0) = g(0)$.

Let (φ, U) be a chart on M at $f(0)$ such that $\varphi(U, f(0)) = (\mathbb{R}^n, 0)$. We can radially isotop f and g to embeddings in U; consider, for example, the isotopy

$$(x,t) \mapsto f((1 - t + t\varepsilon)x), \qquad x \in D^k, \qquad 0 \leqslant t \leqslant 1,$$

for sufficiently small $\varepsilon > 0$. Therefore we assume

$$f(D^k) \cup g(D^k) \subset U.$$

If $k = n$ we can further assume that f and g both preserve or both reverse orientations, as embeddings into the orientable manifold U. If M is orientable this follows from the hypothesis. If M is not orientable we can replace f, if necessary, by an isotopic embedding obtained by isotoping f around an orientation reversing loop based at $f(0)$.

It suffices to show that φf, $\varphi g: D^k \to \mathbb{R}^n$ are isotopic. If $k = n$ we can assume, by proper choice of φ, that both embeddings preserve orientation.

Moreover we can also assume, for any k, that φf and φg are *linear*: any embedding $h:D^k \to \mathbb{R}^n$ with $h(0) = 0$ is isotopic to a linear one by the standard isotopy (see proof of Theorem 4.5.3):

$$(x,t) \to \begin{cases} t^{-1}h(tx), & 1 \geqslant t > 0 \\ Dh(0)x, & t = 0. \end{cases}$$

If φf and φg are linear, and $k = n$, then their determinants are both positive, by our orientation assumptions. Hence they are restriction of maps in the same component of $GL(n)$. A smooth path in $GL(n)$ provides the required isotopy. If $k < n$, we can first extend f and g to linear automorphisms of $\mathbb{R}^n$ having positive determinants and then use a path in $GL(n)$. This finishes the proof when $\partial M = \varnothing$.

If $\partial M \neq \varnothing$, first isotop f and g into $M - \partial M$ by isotoping M into $M - \partial M$: this is easily accomplished with a collar on ∂M. Then apply the previous constructions to $f, g:D^k \to M - \partial M$.

QED

An argument similar to the proof of Theorem 3.1 applies to embeddings of a disjoint union of disks. The following result about pairs of disks generalizes readily to any number of disks.

3.2. Theorem. *Let M be a connected n-manifold without boundary. Suppose that $f_i, g_i : D^n \to M$ $(i = 1, 2)$ are embeddings such that*

$$f_1(D^n) \cap f_2(D^n) = \varnothing = g_1(D^n) \cap g_2(D^n).$$

If M is orientable suppose further that f_i and g_i both preserve, or both reverse, orientation. Then there is a diffeomorphism $H:M \to M$ which is diffeotopic to the identity such that $Hf_i = g_i$ $(i = 1, 2)$.

Proof. By Theorem 3.1, f_1 and g_1 are ambiently isotopic. Hence there is a diffeomorphism H_1 of M, diffeotopic to the identity, with $H_1 f_1 = g_1$. We now apply Theorem 3.1 the embeddings

$$H_1 f_2, g_2 : D^n \hookrightarrow M^n - g_1(D^n).$$

There is diffeomorphism H_2 of $M^n - g_1(D^n)$ such that $H_2 H_1 f_2 = g_2$, and H_2 is isotopic to the identity by a diffeotopy with compact support. Such a diffeotopy extends to all of M so as to leave $g_1(D^n)$ fixed. Therefore H_2 extends to a diffeomorphism of M which is diffeotopic to 1_M and such that $H_1 g_1 = g_1$. The theorem is proved by setting $H = H_2 H_1$

QED

Finally we consider diffeotopies of the circle.

3.3. Theorem. *Every diffeomorphism of S^1 is isotopic to the identity or to complex conjugation. Therefore every diffeomorphism of S^1 extends to a diffeomorphism of D^2.*

Proof. Let $f: S^1 \to S^1$ be a diffeomorphism. First suppose f has degree 1. By a preliminary isotopy we may assume f is the identity on some open interval $J \subset S^1$. Let $J' \subset S^1$ be an open interval such that $J \cup J' = S^1$. Identify J' with an interval of real numbers. An isotopy from f to the identity is given by

$$f_t(x) = \begin{cases} x & \text{if} \quad x \in J \\ tx + (1-t)f(x) & \text{if} \quad x \in J'. \end{cases}$$

Now suppose $\deg f = -1$. Let $\delta: S^1 \to S^1$ be complex conjugation. Then $\deg(f\delta) = 1$ so $f\delta$ is isotopic to the identity by an isotopy g_t. Then $g_t\delta$ is an isotopy from f to δ.

QED

3.4. Corollary. *Let M be a compact 2-manifold without boundary admitting a Morse function having only 2 critical points. Then $M \approx S^2$.*

Proof. By Theorem 5.2.5 (and its proof) M is the union of two 2-disks glued along their boundaries. We may take the disks to be the upper and lower hemispheres of S^2; by Theorem 3.3 we may take the gluing map to be the identity. The result now follows from Theorem 2.2.

Exercises

1. An embedding $f: S^{k-1} \hookrightarrow M$ is *unknotted* if f extends to an embedding of D^k.

(a) f is unknotted if and only if there is a chart $\varphi: U \to \mathbb{R}^n$ on M and an isotopy F of f such that $\varphi F_1: S^{k-1} \hookrightarrow \mathbb{R}^n$ is the standard inclusion.

(b) An embedding $f: S^1 \hookrightarrow \mathbb{R}^3$ is unknotted if and only if there is a compactly supported isotopy of $\mathbb{R}^3$ carrying f to the standard inclusion.

(c) Let M be a simply connected 4-manifold. Then every embedding $S^1 \hookrightarrow M$ is unknotted.

2. The orthogonal group $O(n)$ is a deformation retract of $\mathrm{Diff}^r_W(\mathbb{R}^n)$, $1 \leqslant r \leqslant \omega$.

3. If M is an orientable manifold denote by $\mathrm{Diff}_+(M)$ the group of orientation preserving diffeomorphisms. Let $G \subset \mathrm{Diff}_+(S^n)$ be the image of $\mathrm{Diff}_+(D^{n+1})$ under the restriction homomorphism.

(a) If $f \in \mathrm{Diff}_+(S^n)$ is isotopic to the identity, then $f \in G$;

(b) Let $g, h \in \mathrm{Diff}_+(S^n)$. Then g and h are isotopic to diffeomorphisms u, v which are the identity on the upper and lower hemispheres, respectively; this implies $uv = vu$.

(c) The quotient group $\Gamma_{n+1} = \mathrm{Diff}_+(S^n)/G$ is abelian.

(d) $\Gamma_2 = \{0\}$.

**(e) $\Gamma_3 = \{0\}$. (Smale [2]; Munkres [2].) [Hint: Let $f \in \mathrm{Diff}_+(S^2)$. By an isotopy assume f is the identity on a hemisphere. It now suffices to show that a diffeomorphism g of $\mathbb{R}^2$, having compact support $K \subset \mathbb{R}^2$, is isotopic to the identity through such diffeomorphisms. The unit tangents to images of horizontal lines form a vector field X on $\mathbb{R}^2$ which is constant outside K, and X is homotopic to a constant rel $\mathbb{R}^2 - K$. By the Poincaré–Bendixson theorem, such a homotopy gives rise to an isotopy of g.]

Remark. These groups Γ_i are important in classifying differential structures. The set of diffeomorphism classes of oriented differential structures on S^i forms a group under connected sum. This group is isomorphic to Γ_i except perhaps for $i = 4$. It is known that the Γ_i are finite for all i. The first nontrivial group is $\Gamma_7 \approx \mathbb{Z}_{28}$. For an interesting (and difficult!) Morse-theoretic proof that $\Gamma_4 = \{0\}$ see J. Cerf [1].

Chapter 9

Surfaces

Un des problèmes centraux posés à l'esprit humain est le problème de la succession des formes.

—R. Thom, *Stabilité Structurelle et Morphogénèse*, 1972

Concerned with forms we gain a healthy disrespect for their authority . . .

—M. Shub, For Ralph, 1969

A *surface* is a two-dimensional manifold. The classification of compact surfaces was "known," in some sense, by the end of the nineteenth century. Möbius [1] and Jordan [1] offered proofs (for orientable surfaces in $\mathbb{R}^3$) in the 1860's. Möbius' paper is quite interesting; in fact he used a Morse-theoretic approach similar to the one presented in this chapter. The main interest in Jordan's attempt is in showing how the work of an outstanding mathematician can appear nonsensical a century later.

Of course in those days very few topological concepts had been developed. Both Jordan and Möbius considered two surfaces equivalent if they could be "decomposed into infinitely small pieces in such a way that contiguous pieces of one correspond to contiguous pieces of the other." The difficulties of trying to prove anything on the basis of such a definition are obvious.

The main idea in classifying surfaces goes back to Riemann: cut the surface along closed curves, and arcs joining boundary points, until any further cuts will disconnect it. The maximal number of cuts which can be made without disconnecting the surface, plus 1, was called the *connectivity* by Riemann. Thus a sphere or disk has connectivity 1, or is *simply connected*; an annulus has connectivity 2; a torus has connectivity 3; and so on. What Möbius and Jordan tried to prove is that two compact connected oriented surfaces are homeomorphic if and only if they have the same connectivity.

Riemann proved, more or less, the subtle fact that every maximal set of non-disconnecting cuts has the same cardinality. It seems strange that neither Riemann nor anyone else in the nineteenth century, except perhaps Möbius, seems to have realized the necessity for proving that the connectivity of a compact surface is in fact *finite*.

If one grants the finiteness of the connectivity, the classification reduces to that of simply connected surfaces. The latter is another deep result, for which the nineteenth century offers little in the way of proof.

The connectivity is an intuitively appealing concept, but perhaps for this very reason, it is hard to work with. It is best treated by means of homology theory (see Exercise 17 of Section 9.3).

It turns out that every compact connected surface M is diffeomorphic to one obtained as follows. Punch out a number v of 2-disks from S^2; glue in g cylinders (if M is orientable) or g Möbius bands (if M is nonorientable). The number g, called the genus of M, is uniquely determined by M. The diffeomorphism class of M is characterized by its genus, orientability, and number of boundary components.

The proof of this classification is structured as follows. In Section 9.1 model surfaces are constructed and analyzed. The hardest step is in Section 9.2: the proof that a surface is a disk if it has an admissible Morse function with 2 minima, 1 saddle and no other critical points. The proof given extends to higher dimensions. The classification is completed in Section 9.3 by induction on the number of saddles of a Morse function on the surface.

1. Models of Surfaces

Here is a way of constructing a surface. Start with a surface M and an embedding

$$f: S^0 \times D^2 \hookrightarrow M - \partial M.$$

The image of f is a pair of disjoint disks in M. Now cut out the interior of these disks and glue in the cylinder $D^1 \times S^1$ by $f|S^0 \times S^1$. This produces a new surface M':

$$M' = [M - \operatorname{Int} f(S^0 \times D^2)] \bigcup\nolimits_f D^1 \times S^1.$$

We give M' a differential structure inducing the original structure on $M - \operatorname{Int} f(S^0 \times D^2)$ and $D^1 \times S^1$. By Theorem 8.2.1, this structure is unique up to diffeomorphism. We shall pretend that M' is a well-defined differentiable manifold and write $M' = M[f]$. We say M' is obtained from M by *attaching a handle*, or by *surgery on* f.

1.1. Theorem. *Let M be a surface and let $f_0, f_1: S^0 \times D^2 \to M - \partial M$ be isotopic embeddings. Then $M[f_0] \approx M[f_1]$.*

Proof. By the isotopy extension Theorem 8.1.3 there is a diffeomorphism $\varphi: M \to M$ such that $\varphi f_0 = f_1$. Put $M - \operatorname{Int} f_i(S^0 \times D^2) = Q_i$, $i = 0, 1$. Put

$$g_i = f_i^{-1}|\partial Q_i: \partial Q_i \approx S^0 \times S^1.$$

Then

$$M[f_0] = (D^1 \times S^1) \bigcup\nolimits_{g_0} Q_0$$
$$M[f_1] = (D^1 \times S^1) \bigcup\nolimits_{g_1} Q_1.$$

The diffeomorphism $g_1^{-1}g_0:\partial Q_0 \to \partial Q_1$ extends to the diffeomorphism $\varphi:Q_0 \approx Q_1$. The theorem follows by Theorem 8.2.2.

QED

1.2. Corollary. *Let M be a connected surface. If M is nonorientable, all surfaces obtained by attaching a handle to M are diffeomorphic.*

Proof. Use Theorem 8.3.2.

QED

We give $S^0 \times D^2$ its product orientation. This means that $1 \times D^2$ is oriented like the standard orientation of D^2, while $(-1) \times D^2$ is given the opposite orientation. This orientation of $S^0 \times D^2$ induces an orientation of $S^0 \times S^1$ which is the same as it receives as $\partial(D^1 \times S^1)$, where D^1 and S^1 are given their standard orientations.

Let M be a surface, and let $f:S^0 \times D^2 \to M$ be an embedding. If M can be oriented so that f preserves orientation, we call f an *orientable* embedding. In all other cases f is *nonorientable*. It is easy to prove:

1.3. Theorem. *$M[f]$ is orientable if and only if f is orientable.*

A connected manifold is called *reversible* if it is orientable and admits an orientation reversing diffeomorphism.

1.4. Theorem. *Let M be a connected surface and $f, g:S^0 \times D^2 \to M - \partial M$ embeddings. Then $M[f] \approx M[g]$ in the following cases:*

(a) *M is nonorientable;*

(b) *M is oriented and f and g both preserve or both reverse orientation;*

(c) *M is reversible and both f and g are orientable.*

Proof. Part (a) has already been proved (Corollary 1.2). Part (b) follows from Theorem 8.3.2. To prove (c) it suffices to consider the case where M is oriented so that f preserves and g reverses orientation [since other cases are covered by (b)]. Let $h:M \to M$ reverse orientation. Then $M[hg] \approx M[f]$ by (b). We must prove $M[hg] \approx M[g]$. Let $\rho:S^0 \times D^2 \to S^0 \times D^2$ be the orientation reversing diffeomorphism $\rho(x,y) = (-x,y)$. Then $M[hg] \approx M[g\rho]$. But since $\rho|S^0 \times S^1$ extends to a diffeomorphism of $D^1 \times S^1$, it follows from Theorem 8.2.2 that $M[g\rho] \approx M[g]$.

QED

1.5. Lemma. *Let M be a reversible surface and $f:S^0 \times D^2 \to M - \partial M$ an orientable embedding. Then $M[f]$ is reversible.*

Proof. Let $h:M \to M$ be an orientation reversing diffeomorphism. Let $\rho:S^0 \times D^2 \to S^0 \times D^2$ be an orientation reversing diffeomorphism such that $\rho|S^0 \times S^1$ extends to a diffeomorphism $\bar{\rho}$ of $D^1 \times S^1$.

By ambient isotopy of disks, there is a diffeotopy of M carrying h to a diffeomorphism $g:M \to M$ such that $gf = f\rho$. Note that g reverses orientation.

Consider the map $\varphi: M[f] \to M[f]$ which is $\bar{p}$ on $D^1 \times S^1$ and g on $M - \operatorname{Int} f(S^0 \times D^2)$. Clearly φ is an orientation reversing homeomorphism. By uniqueness of gluing (Theorem 8.2.1) φ can be made into a diffeomorphism.

QED

We now define an important class of surfaces. Let $p \geqslant 0$ be an integer. An orientable surface M is of *genus p* provided M can be obtained from S^2 by successively attaching handles p times. That is, there must exist a sequence of orientable surfaces $M_0, \ldots, M_p$ and orientable embeddings $f_i: S^0 \times D^2 \to M_{i-1}$, $i = 1, \ldots, p$ (if $p > 0$) such that

$$M_0 \approx S^2, \qquad M_i \approx M_{i-1}[f_i], \qquad M_p = M.$$

Thus each M_i has genus i. Later we shall also define nonorientable surfaces of genus p.

Induction on p shows that *an orientable surface of genus p is compact, connected and reversible* (use Theorems 1.3 and 1.5). *It has Euler characteristic* $2 - 2p$ (use Exercise 7, Section 5.2). Therefore *orientable surfaces of different genus are not diffeomorphic.* On the other hand induction on p and Theorem 1.4(c) shows that *two orientable surfaces of the same genus are diffeomorphic.*

In Section 9.3 we shall prove the main theorem of surface theory: every compact connected orientable surface has a genus.

Starting from two connected surfaces M, N without boundary, we construct the *connected sum* of M and N as follows. Take M and N to be disjoint. Let $f: S^0 \times D^2 \to M \cup N$ be an embedding with $f(1 \times D^2) \subset M$, $f(-1 \times D^2) \subset N$. Let $W = (M \cup N)[f]$.

The diffeomorphism class of W is independent of f provided at least one of M, N is nonorientable or reversible. In such a case we pretend W is a well-defined manifold and write $W = M \# N$.

We can also view $M \# N$ as formed by gluing together $M - \operatorname{Int} B$ and $N - \operatorname{Int} D$ by a diffeomorphism $\partial D \approx \partial B$, where $B \subset M$ and $D \subset N$ are disks.

The connected sum of higher dimensional manifolds can be defined analogously.

Clearly $M_0 \# M_1$ is orientable if and only both M_0 and M_1 are orientable. It is easy to prove that if M_i is an orientable surface of genus p_i then $M_0 \# M_1$ is orientable with genus $p_0 + p_1$. In particular an orientable surface of genus $p \geqslant 2$ is the connected sum p tori (Figure 9–1).

We turn to models of nonorientable surfaces. Let P denote the projective plane. A *nonorientable surface of genus* $p \geqslant 1$ means any surface diffeomorphic to the connected sum of p (disjoint) copies of P. Such a surface is nonorientable.

The *Möbius band B* is the surface which is the quotient space of $S^1 \times [-1,1]$ under the identifications $(x,y) \sim (-x,-y)$. It is the simplest

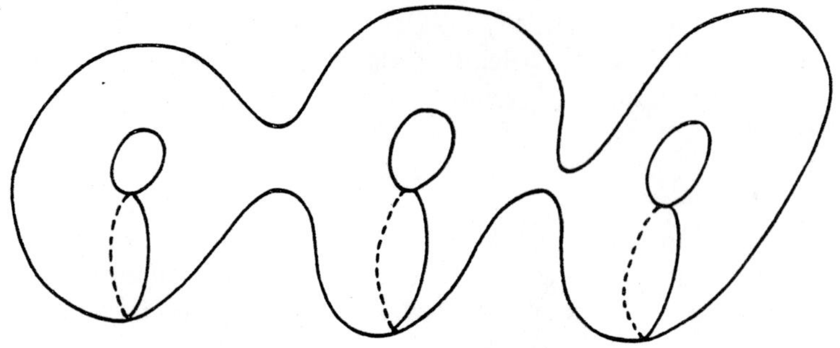

Figure 9–1. Connected sum of 3 tori.

nonorientable surface. Note that $\partial B \approx S^1$. Any surface diffeomorphic to B is also called a Möbius band.

If M is any surface without boundary then

$$M \# P \approx (M - \text{Int } D) \bigcup_f B$$

where $D \subset M$ is a disk and $f: \partial B \approx \partial M$ is arbitrary. This is because the projective plane is obtained by gluing D^2 and B along their boundaries.

The image of B in $M \# P$ is also called a *crosscap*, especially when we think of it as obtained as above by gluing. A nonorientable surface of genus $p \geq 1$ is also called a sphere with p crosscaps attached.

An easy computation shows that the Euler characteristic of a nonorientable surface of genus p is $2 - p$.

1.6. Theorem. *Let M, N be nonorientable surfaces of genus p, q respectively. Then $M \approx N$ if and only if $p = q$.*

Proof. Left as an exercise.

A nonorientable surface of genus 2 is called a *Klein bottle*. It can be obtained from a sphere by attaching a handle by any nonorientable embedding $S^0 \times D^2 \to S^2$.

Let M be a connected nonorientable surface without boundary. Let $f: S^0 \times D^1 \hookrightarrow M$ and consider $M[f]$. We may suppose the image of f is contained in the interior of a small disk $D \subset M$. Thus $M[f]$ is obtained by gluing together $M - \text{Int } D$ and $D[f]$ along their boundaries. If we identify D with a hemisphere of S^2, and reinterpret f as an embedding $g: S^0 \times D^2 \to S^2$, we find that $M[f] \approx M \# S^2[g]$. Now $S^2[g]$ is a torus if g is orientable, and Klein bottle otherwise; and the orientability of g is the same as that of $f: S^0 \times D^2 \hookrightarrow D$. But since M is nonorientable, we can isotop $f(1 \times D^2)$ around an orientation reversing loop. This leads to another embedding $f_1: S^0 \times D^2 \to D$ which is isotopic to f in M, and which is orientable if and only if f is nonorientable. Thus

$$M[f] \approx M \# S^2[f] \approx M \# S^2[f_1]$$

where $S^2[f]$ is a torus and $S^2[f_1]$ is a Klein bottle. Since a Klein bottle is a sphere with two crosscaps, this proves:

1.7. Theorem. *Attaching a handle to a connected nonorientable surface is the same as attaching two crosscaps. Therefore attaching a handle to a nonorientable surface of genus p produces a nonorientable surface of genus $p + 2$.*

A dual result is:

1.8. Theorem. *Let M be an orientable surface of genus p. Attaching a crosscap to M yields a nonorientable surface of genus $2p + 1$.*

Proof. This is clear if $p = 0$. If $p > 0$ consider M as the connected sum of p tori. Then $M \# P$ is the same as P with p handles, and the preceding result applies.

QED

We now construct models of ∂-surfaces by simply cutting out the interiors of a number $k > 0$ of disjoint disks from an orientable or nonorientable surface M of genus g. The result is called a ∂-surface of genus g with k boundary components. By isotopy of disks (Section 8.3), the diffeomorphism class of such a surface depends only on M and the number of disks.

By a *model surface* we mean a surface or ∂-surface of genus p, orientable or nonorientable. In Section 9.3 we shall show that every compact connected surface is diffeomorphic to a unique type of model surface.

It is clear that two model surfaces are diffeomorphic if and only if (a) they have the same genus and the same number of boundary circles, and (b) both are orientable or both are nonorientable.

If a model surface M has genus g, and ∂M has b components, then its Euler characteristic χ is $2 - 2g - b$ if M is orientable, while $\chi = 2 - g - b$ if M is not orientable. (See Exercise 7, Section 5.2) This proves:

1.9. Theorem. *Two model surfaces are diffeomorphic if and only if they have the same genus, the same Euler characteristic, and the same number of boundary components.*

Exercises

1. An orientable surface of genus p contains p disjoint circles whose union does not separate the surface.

2. A nonorientable surface of genus p contains p disjoint circles each of which reverses orientation.

3. Let $C \subset M$ be a circle in a surface M without boundary, which does not disconnect M.

(a) If C reverses orientation, it has a Möbius band neighborhood and there is a surface N such that $M \approx N \# P^2$.

(b) If M is orientable there is a circle $C' \subset M$ meeting C transversely, and at only one point. Moreover:

(c) $C \cup C'$ has a neighborhood $N \subset M$ diffeomorphic to $T - \operatorname{Int} D$ where D is a disk in the torus T. Consequently $M \approx W \# T^2$ for some surface W.

4. (a) Every orientable surface of genus p bounds a compact 3-manifold.

(b) A nonorientable surface of genus p bounds a compact 3-manifold if and only if p is even. [Use Exercise 7, Section 5.2.]

5. The complex projective plane is not reversible. [See Exercise 18, Section 5.2.]

***6. Is every orientable 3-manifold reversible? (Perhaps Lickorish [1] is useful here.)

7. Let M be an n-manifold without boundary. Then $M \# S^n \approx M$.

8. Let $f: S^1 \to M$ be a loop in a surface M. Then f preserves orientation (in the sense of Section 4.4) if and only if $\#_2(f,f) = 0$ (see Exercise 4, Section 5.2 for the mod 2 intersection number $\#_2$).

2. Characterization of the Disk

The following result is the key to the classification of surfaces.

2.1. Theorem. *Let $f: M \to \mathbb{R}$ be an admissible Morse function on a compact connected surface M. Suppose f has exactly 3 critical points, and these are of type 0, 0, 1. Then $M \approx D^2$.*

The strategy of proof is as follows. First we find another function $g: N \to \mathbb{R}$ of the same kind, on a surface N which we know is diffeomorphic to D^2. Then we construct a homeomorphism from M to N using level curves and gradient lines of the two Morse functions. This homeomorphism is then smoothed to a diffeomorphism.

Before beginning the proof we discuss a method of extending diffeomorphism.

Let M_i be a complete Riemannian manifold, $i = 0, 1$, and $f_i: M_i \to \mathbb{R}$ a map. For $x \in M_i$ let $\Lambda_i(x) \subset M_i$ be the maximal solution curve through x of the vector field grad f_i.

Let $U_i \subset M_i$ be open and let $G: U_0 \approx U_1$ be a diffeomorphism having the following properties: for all $x \in U_0$,

$$f_1 G(x) = f_0(x), \text{ and} \tag{1}$$

$$G(U_0 \cap \Lambda_0(x)) = U_1 \cap \Lambda_1(Gx). \tag{2}$$

We say that G *preserves level surfaces and gradient lines.*

Let $U_i^* \subset M_i$ be the *saturation* of U_i, under the flow of grad f_i, that is:

$$U_i^* = \bigcup_{x \in U_i} \Lambda_i(x).$$

2.2. Lemma. *In addition to the above, suppose also that for each $x \in U_0$,*
(a) $f_0(\Lambda_0(x)) = f_1(\Lambda_1(f(x)))$;
(b) $\Lambda_0(x) \cap U_0$ is connected.

Then G extends to a unique diffeomorphism $G = U_0^* \approx U_1^*$ which satisfies (1) and (2).

Proof. Any critical points of f_0 in U_0^* are already in U; thus G is already defined in a neighborhood of such points. If $x \in U_0^* - U_0$, then $\Lambda_0(x)$

contains a point $y \in U_0$. Define

$$G(x) = \Lambda_1(Gy) \cap f_1^{-1}(f_0(x)).$$

The intersection is nonempty by (a); it contains only one point since f_1 is monotone on gradient lines; and $G(x)$ is independent of y by (b).

It remains to prove that G is C^∞. Let $x \in U_0^*$; put $G(x) = y$. Then from commutativity of the diagram

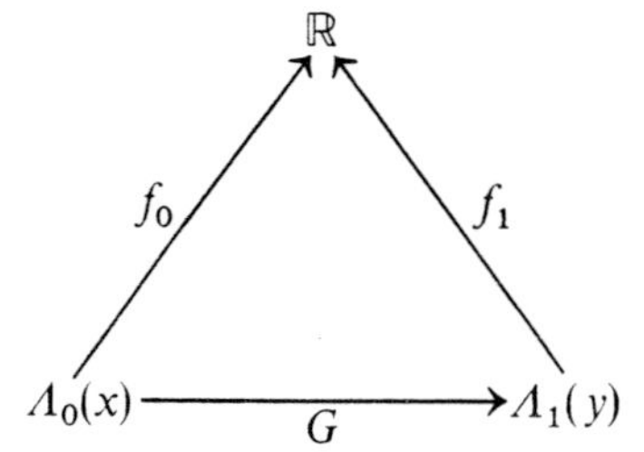

we see that

$$G|\Lambda_0(x) = [f_1|\Lambda_1(y)]^{-1} \circ [f_0|\Lambda_0(x)].$$

Thus $G|\Lambda_0(x)$ is C^∞. That G is globally C^∞ follows from the fact that the gradient flows are C^∞.

QED

Notice that Theorem 2.1 is also true if U_0 and U_1 are open subsets of level surfaces ((b) is then trivial). The proof is the same.

Next we study the model function

$$g: \mathbb{R}^2 \to \mathbb{R}$$

$$g(x,y) = \frac{x^4}{4} - \frac{x^3}{3} - x^2 + y^2$$

$$= \int_0^x t(t + 1)(t - 2)\, dt + y^2.$$

The critical points of g are at $(0,0)$, $(-1,0)$, $(2,0)$. They are nondegenerate, of types 1, 0, 0 respectively. Note that the three critical values are distinct.

2.3. Lemma. *$g^{-1}(c)$ is connected if $c \geqslant 0$.*

Proof. Note that g is proper, so $g^{-1}(c)$ is compact. One component of the critical level $g^{-1}(0)$ is a figure 8, and each loop of the figure 8 encloses one of the two (local) minima of g. Any other component of $g^{-1}(0)$ would be a circle enclosing another minimum, which is impossible. Therefore $g^{-1}(c)$ is connected. If $c > 0$, each component of $g^{-1}(c)$ is a circle enclosing a minimum. But one component encloses $g^{-1}(0)$, hence it encloses both minima. Therefore $g^{-1}(c)$ has only one component.

QED

2.4. Lemma. *If $\xi > 0$, $g^{-1}(-\infty,\xi] \approx D^2$.*

Proof. It suffices to prove this for some large ξ, by the regular interval Theorem 5.2.3. We shall show that if α is large then each ray from (0,0) meets the curve $g^{-1}(\xi)$ transversely. Since $g^{-1}(\xi)$ is connected, this will mean $g^{-1}(-\infty,\xi]$ is star shaped, and thus prove the lemma. Since grad f is perpendicular to $g^{-1}(\xi)$, it suffices to show that if $|x|^2 + |y|^2$ is large enough then

$$\langle \text{grad } f(x,y), (x,y)\rangle \neq 0$$

But

$$\begin{aligned}\langle \text{grad } f(x,y), (x,y)\rangle &= \langle (x^3 - x^2 - 2x, 2y), (x,y)\rangle \\ &= x^2(x^2 - x - 2) + 2y^2\end{aligned}$$

which is positive if $x > 2$ or $y > 3$.

QED

Now consider the gradient flow Φ_t of g, given by the system of differential equations

$$\frac{dx}{dt} = x^3 - x^2 - 2x$$

$$\frac{dy}{dt} = 2y.$$

These can be easily solved (see Figure 9–2). It is clear that the x-axis and y-axis are invariant. The stationary points of the flow are of course the critical points of g. There are two sources, $(-1,0)$ and $(2,0)$; and one saddle, (0,0). The flow lines are orthogonal to the level curves $g =$ constant.

If $y \neq 0$, or $y = 0$ and $x < 1$ or $x > 2$, then $|\Phi_t(x,y)| \to \infty$ as $t \to \infty$. If $-1 \leqslant x < 0$ or $0 < x \leqslant 2$ then $\Phi_t(x,0) \to (0,0)$ as $t \to \infty$.

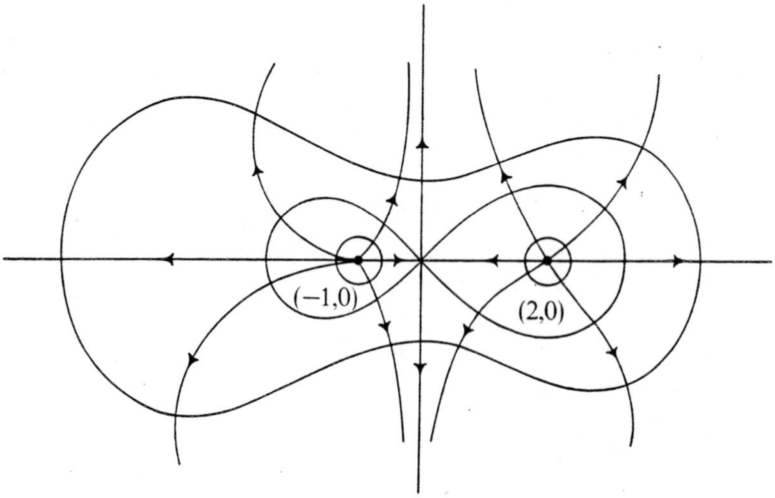

Figure 9–2. Levels and gradients.

As $t \to -\infty$, $\Phi_t(x,y) \to (-1,0)$ if $x < 0$; to $(0,0)$ if $x = 0$; and to $(2,0)$ if $x > 0$.

Now let $f: M \to \mathbb{R}$ be as in Theorem 2.1. Let $a, b, c \in M$ be the three critical points of f, with a and c local minima and b a saddle. By slightly perturbing f near a and c, if necessary, we can assume that $f(b) > f(a) > f(c)$.

Let $\lambda: \mathbb{R} \to \mathbb{R}$ be a diffeomorphism such that

$$\begin{aligned} \lambda(f(a)) &= -\tfrac{11}{12} = g(-1,0), \\ \lambda(f(b)) &= 0 = g(0,0), \\ \lambda(f(c)) &= -\tfrac{8}{3} = g(2,0). \end{aligned}$$

The map $\lambda \circ f: M \to \mathbb{R}$ is a Morse function having the properties listed in Theorem 2.1. Therefore we may assume that $f(a) = g(-1,0)$, $f(b) = g(0,0)$, $f(c) = g(2,0)$.

Set $\xi = f(\partial M) > 0$. Then ξ is the maximum value of f, and $f^{-1}(\xi) = \partial M$.

We first prove that ∂M is connected. By the regular interval Theorem 6.2.2 it suffices to prove that $f^{-1}(\varepsilon)$ is connected for some ε, $0 < \varepsilon < \xi$, since $\partial M \approx f^{-1}(\varepsilon)$.

Give M a Riemannian metric induced by Morse charts near critical points. Let F_t be the flow of $-\operatorname{grad} f$; then $F_t: M \to M$ is defined for all $t \geqslant 0$. For each $x \in M$ the limit $\bar{x} = \lim_{t \to \infty} F_t(x)$ is one of the three critical points. The sets

$$\begin{aligned} W_a &= \{x \in M : \bar{x} = a\}, \\ W_c &= \{x \in M : \bar{x} = c\} \end{aligned}$$

are disjoint open sets.

A glance at a Morse chart near the saddle b shows there are only two nonconstant trajectories limiting at b, and these intersect $f^{-1}(\varepsilon)$ in two points, say q_1, q_2. Moreover the set $\{q_1, q_2\}$ is the common boundary in $f^{-1}(\varepsilon)$ of $f^{-1}(\varepsilon) \cap W_a$ and $f^{-1}(\varepsilon) \cap W_b$. No component of $f^{-1}(\varepsilon)$ can be entirely in W_a, for if it were, W_a would be a component of M, contradicting connectedness of M. Similarly for W_b. Therefore each component of $f^{-1}(\varepsilon)$ must be separated by a subset of $\{q_1, q_2\}$. Since a single point cannot separate, each component contains both q_1 and q_2. Hence there is only one component.

Let $\varphi: U \to \mathbb{R}^2$ and $\psi: V \to \mathbb{R}^2$ be Morse charts for f and g at b and $(0,0)$. We may assume that $\varphi(U) = \psi(V)$. Put

$$H = \varphi^{-1}\psi: V \to M.$$

Then H maps V diffeomorphically onto U, preserving level surfaces (i.e., $gH = f$). We can choose V so that M has a Riemannian metric making H an isometry (where V inherits the standard metric from $\mathbb{R}^2$). Then H also preserves gradient lines.

We also choose V so that V meets each g-gradient line in a connected set; for example, $V = \operatorname{Int} B_\delta$ for small $\delta > 0$. It then follows from Theorem 2.2 that H extends to a diffeomorphism $H: V^* \approx W^*$ between the saturation of V and W, preserving levels and gradients.

We can choose ψ so that H has the following property: as $t \to -\infty$, the grad f trajectory of $H(p)$ tends to a or c, respectively, according as the grad g trajectory of $p \in V^*$ tends to $(-1,0)$ or $(2,0)$.

Fix a real number α such that

$$g(2,0) < g(-1,0) < \alpha < g(0,0) = 0.$$

Let D, D' be the components of the saddles $(-1,0)$, $(2,0)$ respectively, in the submanifold $g^{-1}(-\infty,\alpha] \subset \mathbb{R}^2$. Morse's lemma implies that D and D' are disks. Observe that ∂D and $\partial D'$ are components of level curves.

Let $\varepsilon > 0$ be very small. Let $B_\varepsilon \subset \mathbb{R}^2$ be the square $|x| \leqslant \varepsilon$, $|y| \leqslant \varepsilon$. Let B_ε^* denote the saturation of B_ε under the gradient flow of g.

Define

$$P_\varepsilon = [B_\varepsilon^* - \operatorname{Int}(D \cup D')] \cap g^{-1}(-\infty,\varepsilon].$$

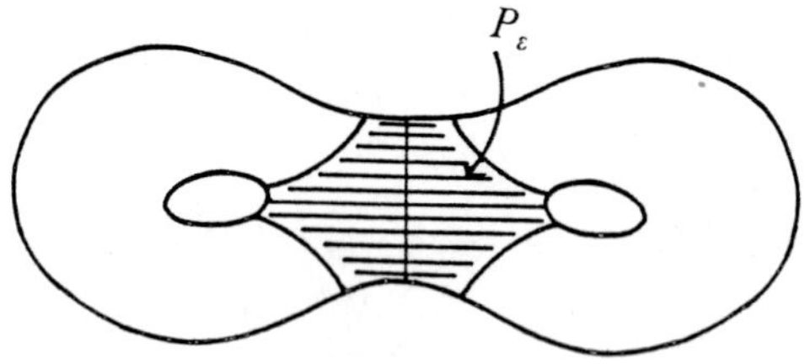

Figure 9–3.

See Figure 9–3. If ε is sufficiently small then $B_\varepsilon \subset V$ and $P_\varepsilon \subset V^*$; and also the sets

$$A = P_\varepsilon \cap D = P_\varepsilon \cap \partial D,$$
$$A' = P_\varepsilon \cap D' = P_\varepsilon \cap \partial D'$$

are (compact) arcs. Fix such an ε.

The diffeomorphism $H: V^* \approx W^*$ embeds the arcs A, A' in the circles $\partial\Gamma$, $\partial\Gamma'$ respectively. *There are diffeomorphisms $D \approx \Gamma$, $D' \approx \Gamma'$ which agree with H on A, A.* To see this, identify D and Γ with D^2 via diffeomorphisms in such a way that H/A is transformed into an orientation preserving embedding H_0 of an arc $B \subset \partial D^2$ into ∂D^2. By isotopy of disks, H_0 is isotopic to the inclusion of B in ∂D^2, and by isotopy extension H_0 therefore extends to a diffeomorphism of D^2. Hence H extends to a diffeomorphism $D \approx \Gamma$, and likewise for D' and Γ'.

Let Γ, Γ' be the components of a and c respectively in $f^{-1}(\infty,\alpha]$. Then Γ and Γ' are disjoint disks.

In this way we obtain a diffeomorphism

$$G: D \cup D' \approx \Gamma \cup \Gamma'$$
$$G(D) = \Gamma, \qquad G(D') = \Gamma'.$$

We now extend the restriction

$$G: \partial D \cup \partial D' \approx \partial\Gamma \cup \partial\Gamma'$$

to a diffeomorphism of saturations

$$F:(\partial D \cup \partial D')^* \to (\partial\Gamma \cup \partial\Gamma')^*,$$

such that F preserves level curves and gradient lines. For this we use Theorem 2.1 and the remark following its proof.

Notice that F and H agree on

$$(\partial D \cup \partial D')^* \cap P_\varepsilon$$

since they agree on $A \cup A'$, and both preserve level curves and gradients.

Define a map

$$K:g^{-1}(-\infty,\xi] \to M$$

by

$$K = \begin{cases} G & \text{in} \quad D \cup D' \\ H & \text{in} \quad P_\varepsilon \\ F & \text{elsewhere.} \end{cases}$$

Then K is well defined. It is easy to see that K is surjective and injective, hence K is a homeomorphism. Moreover K maps $D \cup D'$ and $g^{-1}(-\infty,\xi] - \operatorname{Int}(D \cup D')$ diffeomorphically.

From the smoothing Theorem 8.1.9 we conclude that $N \approx M$. Since also $N \approx D^2$ by Theorem 2.4 the proof of Theorem 2.1 is complete.

QED

In the proof of Theorem 2.1 we did not use Theorems 8.3.3 and 8.3.4, or any other special properties of manifolds of dimension 1 or 2. The same argument, with only notational changes, proves the following generalization of Theorem 2.1:

2.4. Theorem. *Let $f:M \to \mathbb{R}$ be an admissible Morse function on a compact connected n-manifold M. Suppose f has exactly 3 critical points, of types* 0, 0, 1. *Then $M \approx D^n$.*

Exercises

1. The proof of Theorem 2.1 can be varied slightly to prove the following. Let $f:M \to \mathbb{R}$, $g:D^2 \to \mathbb{R}$ be admissible Morse each having only three critical points, and these types 0, 0, 1. Suppose that f and g take the same values at corresponding critical points, and that $f(\partial M) = g(\partial D^2)$. Then there is homeomorphism $h:D^2 \to M$ such that $fh = g$. Moreover M has a Riemannian metric for which h maps gradient lines of g to gradient lines of f.

2. Let $a < 0 < b$. Define a polynomial map in two variables:

$$P_{a,b}:\mathbb{R} \times \mathbb{R}^{k-1} \to \mathbb{R},$$

$$P_{a,b}(x,y) = \int_0^x s(s-a)(s-b)\,ds + |y|^2.$$

(a) $P_{a,b}$ is a Morse function having local minima at $(a,0)$ and $(b,0)$, a type 1 saddle at $(0,0)$, and no other critical points.

(b) $P_{a,b}(a,0) \neq P_{a,b}(b,0)$ if and only if $a \neq -b$.
(c) Let α be in the image of $P_{a,b}$. Then $P_{a,b}^{-1}(-\infty,\alpha]$ is connected if and only if $\alpha \geqslant 0$.
(d) $P_{a,b}^{-1}(-\infty,\alpha] \approx D^k$ if $\alpha > 0$.

3. Let M be a compact connected surface without boundary which admits a Morse function having just four critical points, exactly one of which is a saddle. Then $M \approx S^2$.

4. Let $f: M \to \mathbb{R}$ be a Morse function on a connected compact Riemannian manifold without boundary. Assume the Riemannian metric comes from Morse charts near critical points. Then if f has more than one local minimum, there exist two local minima a, b and a type 1 critical point p, with the following property: one branch of the unstable manifold of p (for grad f) tends to a and the other branch tends to b.

5. [Smale]. Let M be a compact connected manifold. If $\partial M = \varnothing$ then M has a Morse function with only 1 maximum and 1 minimum. [Use Exercise 4 and Theorem 2.4.]

3. The Classification of Compact Surfaces

We begin by investigating neighborhoods of a critical level of an admissible Morse function f on a compact connected surface M. Let $p \in M$ be a saddle (critical point of index 1). Suppose that $f(p) = 0$ and that $\varepsilon > 0$ is such that p is the only critical point in $N = f^{-1}[-\varepsilon,\varepsilon]$. Assume that N is connected. Put

$$C_- = f^{-1}(-\varepsilon)$$
$$C_0 = f^{-1}(0)$$
$$C_+ = f^{-1}(\varepsilon).$$

Then C_- and C_+ are compact 1-manifolds without boundary; $C_- \cup C_+ = \partial N$. Since N is connected, C_0 must be connected. Therefore C_0 is a figure 8.

Here are two examples. In each f has one saddle and one minimum in M.

(1) M is a U-shaped cylinder in $\mathbb{R}^3$ and f is the height function shown in Figure 9–4. (2) M is a Möbius band, f has the level curves shown in Figure 9–5.

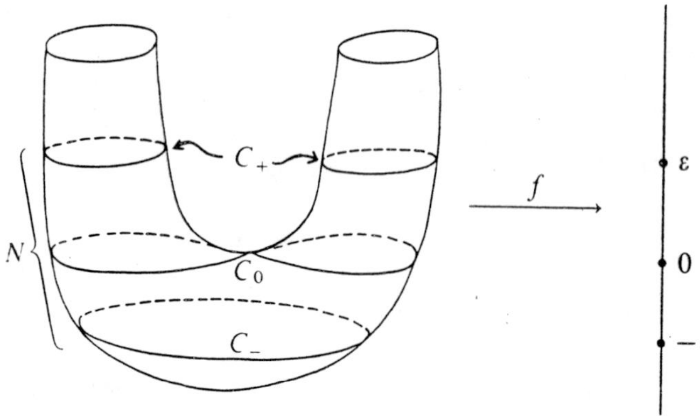

Figure 9–4.

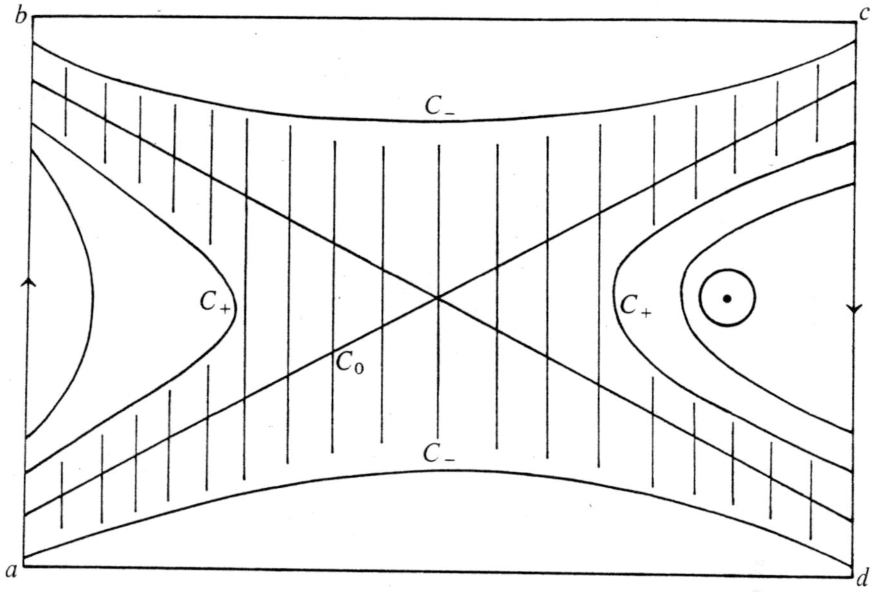

Figure 9–5. *ab* is identified with *cd*. N is shaded.

Note that example (1) is also obtained from Figure 9–5 if *ab* is identified with *dc* to make a cylinder.

In fact these are the only examples of such an N, up to diffeomorphism. We do not need to prove this, but only the following consequences:

3.1. Lemma. *Let f and N be as above. Then either*
(a) *N is orientable and ∂N has three components*; *or*
(b) *N is nonorientable and ∂N has two components.*

Proof. A Morse chart at p shows that p has an X-shaped neighborhood in C_0. Label the four branches of the X by the quadrants they lie in (Figure 9–6). The arrows in Figure 9–6 represent grad f. The key question is: how

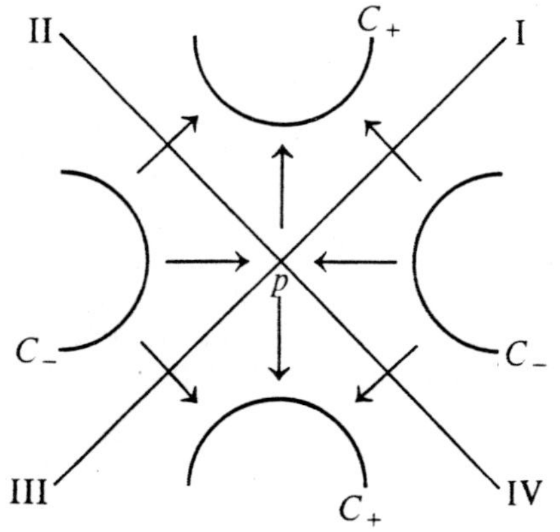

Figure 9–6. Levels and gradients near a saddle.

are the four branches I, II, III, IV connected in C_0? Suppose I connects to IV. The resulting loop λ based at p *preserves orientation*, as can be seen by considering the orientation defined by the tangent to the loop and grad f. (Perturb the loop slightly away from p to make it smoothly embedded. See Figure 9–7.) It is clear that in this case II must connect to III. Notice that

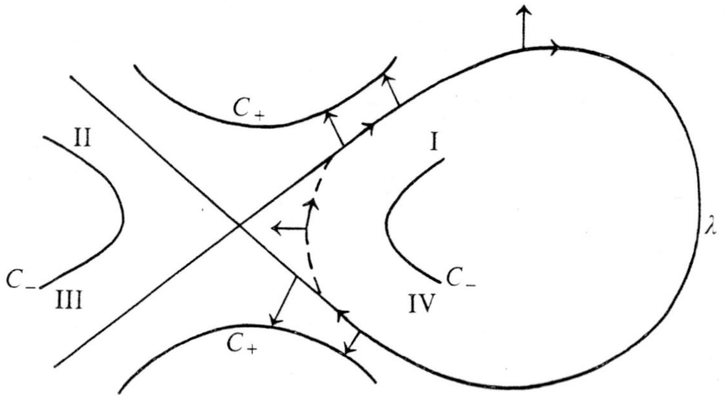

Figure 9–7. The loop λ preserves orientation.

C_+ is connected, for otherwise we could follow the top part of C_+ around the loop and some gradient line would intersect C_+ twice. This is impossible because $C_+ = f^{-1}(\varepsilon)$. (See Figure 9–8.)

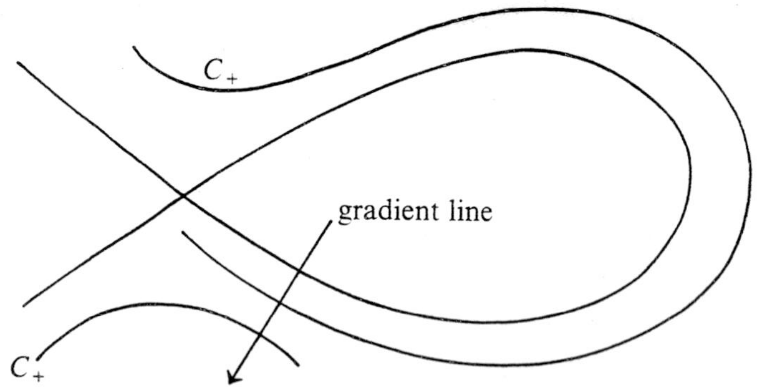

Figure 9–8. Impossible, because $C_+ = f^{-1}(\varepsilon)$.

For similar reasons, the right and left branches of C_- must each close up, forming two components. If I connects to II then C_- is connected and C_+ has two components.

Now suppose I connects to III. The resulting loop then *reverses* orientation (Figure 9–9). In this case the two branches of C_+ connect up; C_+ is

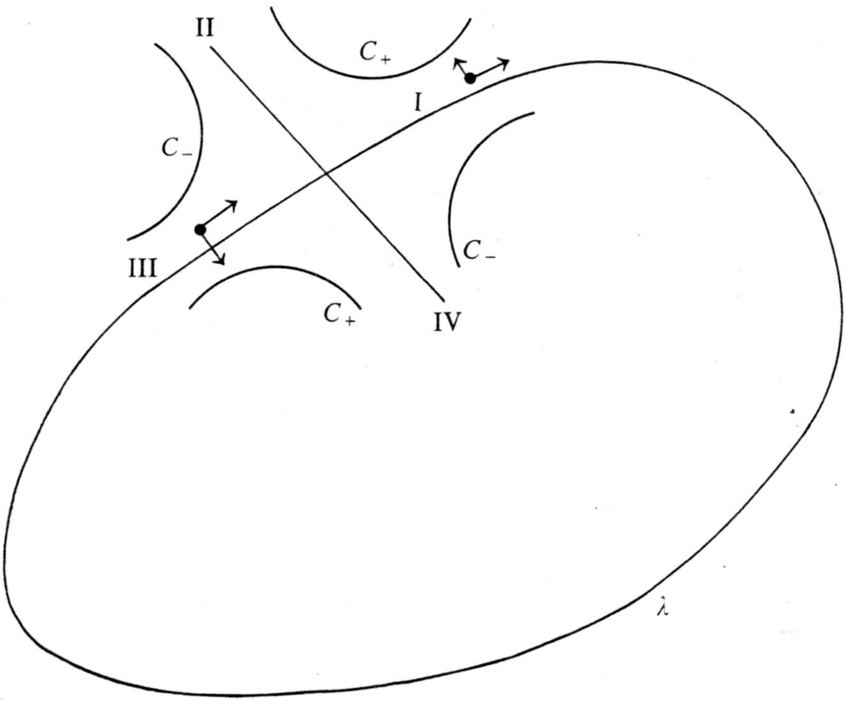

Figure 9–9. λ reverses orientation.

connected. Likewise, II connects to IV and C_- is connected.

QED

Let $B_1, \ldots, B_k \subset D^2 - \partial D^2$ be disjoint embedded disks. Put $H_k = D^2 - \cup \operatorname{Int} B_i$, and $H_0 = D^2$.

A *disk with k holes* means a surface diffeomorphic to H_k, that is, an orientable model surface of genus 0 and $k + 1$ boundary circles. Any two such surfaces (for the same k) are diffeomorphic. Since the Euler characteristic of H_k is $1 - k$, H_k is not diffeomorphic to H_ℓ for $k \neq \ell$. Note that $H_1 \approx S^1 \times I$.

Returning to the situation of Theorem 3.1, we have:

3.2. Lemma. *Let N be as in Theorem 3.1(a). Then N is a disk with 2 holes.*

Proof. We may assume C_- has two components and C_+ has one. Glue disks onto N along each of the components of C_-, to obtain a new manifold V. Define a map $g: V \to \mathbb{R}$ which is f on N, and an each disk is $x^2 + y^2 - 1 - \varepsilon$, when the disk is identified with D^2. The differential structure on V can be chosen so that g is C^∞. (Use collars determined by level curves and gradient lines.) Then g is an admissible Morse function on V having one saddle and two minima. By Theorem 2.1, $V \approx D^2$. Therefore $N \approx H_2$.

QED

A *saddle* is a critical point of index 1, we restate Theorem 3.2 as:

3.3. Theorem. *Let M be a compact connected orientable surface admitting a Morse function having only one critical point, a saddle. There is a disk with 2 holes; moreover f takes its maximum and minimum on ∂M.*

It is now easy to classify compact surfaces admitting a Morse function with only one saddle:

3.4. Theorem. *Let $f:M \to \mathbb{R}$ be an admissible Morse function on a compact connected orientable surface. Suppose f has exactly one saddle (and perhaps other critical points of type 0 or 2). Then M is diffeomorphic to either S^2, D^2, $S^1 \times I$ or H_2. If $f|\partial M$ is constant then $M \not\approx H_2$.*

Proof. Delete from M the interior of disjoint disks around the critical points (if any) of types 0 and 2. Do this in such a way that the disks contain no other critical points, and their boundaries are components of level curves. The resulting manifold W is diffeomorphic to H_2 by Theorem 3.3. If f has no critical points of type 0 or 2, then $M = W$; however, this cannot happen if $f|\partial M$ is constant. If there are critical points of types 0 or 2, then M is obtained from W by capping some of the boundary circles with disks. This produces $S^1 \times I$, D^2 or S^2.

QED

We now come to the *classification of compact orientable surfaces.* First we assume no boundary.

3.5. Theorem. *Let M be a compact connected orientable surface without boundary. Then there is a unique integer $p \geqslant 0$ such that M is an orientable surface of genus p as defined in Section 9.1 (a "sphere with p handles"). The Euler characteristic of M determines p by the formula $\chi(M) = 2 - 2p$. In particular $\chi(M)$ is even and $\leqslant 2$.*

Proof. We proceed by induction on the number v of saddles of a Morse function $f:M \to \mathbb{R}$.

Suppose f has no saddles. Give M a Riemannian metric. Let $P \subset M$ be the set of minima. Each trajectory of grad f in $M - P$ tends toward a maximum. The basin of attraction of each maximum is an open set; but it is also closed since different basins are disjoint. Since $M - P$ is connected, there is only one maximum. Similarly there is only one minimum. Hence $M \approx S^2$ by Theorem 8.3.4.

Let $v = k > 0$ and suppose inductively that the theorem is true whenever M admits a Morse function having fewer then k saddles. We may assume f separates critical points; there is then a unique saddle p such that $f(p) < f(q)$ for every saddle $q \neq p$.

Let $f(p) = \alpha$ and let $\beta > \alpha$ be such that p is the only critical point in $f^{-1}[\alpha,\beta]$.

Let V be the component of p in $f^{-1}(-\infty,\beta]$. Notice that $\partial V = f^{-1}(\beta)$. Since $f|V$ has only one saddle, we can apply Theorem 3.4. Since $\partial V \neq \emptyset$, we conclude that $V \not\approx S^2$. Also $V \not\approx H_2$ since $f|\partial V$ is constant. Thus $V \approx D^2$ or $V \approx S^1 \times I$.

Suppose $V \approx D^2$. Then we can define a new Morse function $g: M \to \mathbb{R}$, equal to f on $M - V$, and having only one critical point in V (a minimum). Since g has only $k - 1$ saddles, it follows from the induction hypothesis that the theorem is true for M.

Suppose finally that $V \approx S^1 \times I$. Then $\partial V = S^1 \times \{0,1\}$. Let M_0 be obtained from M by capping ∂V with two disks. We can redefine f on the disks to get a Morse function $f_0: M_0 \to \mathbb{R}$ having fewer saddles than f. By the induction hypothesis M_0 is an orientable surface of some genus q. Since M is evidently obtained from M_0 by attaching a handle, M has genus $q + 1$. This completes the induction.

The uniqueness of the genus and the formula for the Euler characteristic were proved in Section 9.1.

QED

It is now easy to give a geometric interpretation to the genus.

3.6. Theorem. *Let M be an orientable surface of genus p. Then there exist p disjoint circles in M whose complement is connected; but any $p + 1$ disjoint circles disconnect M.*

Proof. If $p = 0$ we may assume $M = S^2$. The first part of the conclusion is vacuous and the second follows from Theorem 4.4.6.

Suppose $C_1, \ldots, C_q$ are disjoint circles in M, $q \geqslant 1$, and $M - \cup C_i$ is connected. Let $N_1, \ldots, N_q$ be disjoint closed tubular neighborhoods of $C_1, \ldots, C_q$. Let $V = M - \cup \operatorname{Int} N_i$. Let W be obtained from V by capping the $2q$ boundary circles of V with disks. Notice that W is connected and orientable, and that M is obtained from W by attaching q handles. Let W have genus $g \geqslant 0$. Then M has genus $g + q = p$. It follows that $q \leqslant p$.

QED

The classification of compact orientable ∂-surfaces is as follows:

3.7. Theorem. *Let M be a connected compact orientable surface of Euler characteristic χ. Suppose ∂M has $k > 0$ boundary components. Then $\chi + k$ is even. Let $p = 1 - (\chi + k)/2$. Then M is diffeomorphic to the surface obtained from an orientable surface of genus p by removal of the interiors of k disjoint disks.*

Proof. Cap the boundary circles of M to produce an orientable surface W. Then the Euler characteristic $\chi + k$. Therefore if W has genus p, we find that $\chi + k = 2 - 2p$.

QED

The number p associated to the ∂-surface M by Theorem 3.7 is called the *genus of* M.

We turn to nonorientable surfaces.

3.8. Lemma. *Every nonorientable surface N contains a submanifold which is a Möbius band.*

Proof. Pick some orientation reversing loop $f: S^1 \to N - \partial N$. We can assume f is an immersion with clean double points. Thus S^1 breaks up into a finite number of arcs $J_1, \ldots, J_k$, disjoint except for endpoints, each of which is mapped injectively except that the endpoints are identified. We can further homotop each $f_i = f|J_i$ rel ∂J_i so that the tangent vectors at the endpoints coincide. Then each f_i is a loop represented by an embedding $S^1 \to M$. If each f_i preserves orientation so does f. Hence some f_j reverses orientation; a tubular neighborhood of $f_j(J_j)$ is a Möbius band.

QED

3.9. Lemma. *Let N be a compact connected nonorientable surface. Then there is a unique integer $p > 0$ such that M contains p, but not $p + 1$, disjoint Möbius bands.*

Proof. It suffices to exhibit an integer n such that no $n + 1$ Möbius bands in N can be disjoint.

Suppose $B \subset N$ is a Möbius band. Then ∂B is connected; hence $N - B$ is connected. Therefore if $B_1, \ldots, B_p \subset N$ are disjoint Möbius bands, it follows that $N - \cup B_i$ is connected.

Let $\pi: \tilde{N} \to N$ be the orientable double covering of N. Then $\pi^{-1}(B)$ is a cylinder in $\tilde{N}$ if $B \subset N$ is a Möbius band.

Let $V \subset N$ be a connected two-dimensional submanifold. Then $\pi^{-1}(V)$ is connected if and only if V is nonorientable. Hence $\pi^{-1}(V)$ is connected if and only if V contains a Möbius band.

Let the genus of $\tilde{N}$ be $n - 1 \geqslant 0$.

Let $B_1, \ldots, B_n$ be disjoint Möbius bands in N. Then $\pi^{-1}(B_1), \ldots, \pi^{-1}(B_n)$ are disjoint cylinders in $\tilde{N}$; they contain n disjoint embedded circles. Therefore $\tilde{N} - \cup\pi^{-1}(B_i)$, which is $\pi^{-1}(N - \cup B_i)$, is disconnected by Theorem 3.6 It follows that $N - \cup B_i$ is orientable and contains no Möbius band.

QED

We call the integer p of Theorem 3.9 the *Möbius number* of N.

3.10. Theorem. *Let N be a compact connected nonorientable surface without boundary, having Möbius number p. Then N is a nonorientable surface of genus p.*

Proof. Let $M_0 \subset M$ be obtained by cutting out the interiors of p disjoint Möbius bands. Then M_0 is orientable. Cap off ∂M_0 with p disks to obtain an orientable surface M; let M have genus g. Clearly N is formed by attaching p crosscaps to N. By Theorems 1.7 and 1.8 N is a nonorientable surface of genus $2g + p$. But this implies that the Möbius number of N is at least $2g + p$. Therefore $2g = 0$ and M is a sphere.

QED

We conclude with a convenient diffeomorphism criterion:

3.11. Theorem. *Two connected compact surfaces are diffeomorphic if and only if they have the same Euler characteristic and the same number of boundary components.*

Proof. In the orientable case this follows from Theorem 3.7. In the nonorientable case, glue Möbius bands to the boundary components; this preserves the Euler characteristic. The theorem now follows from Theorem 3.10.

Exercises

1. A compact surface embedded in $\mathbb{R}^2$ is diffeomorphic to D^2 or some H_k.

***2.** Let M be a surface and $C \subset M$ a circle. If C is contractible to a point in M then C bounds a disk in M.

***3.** A connected noncompact simply connected surface M without boundary, is diffeomorphic to $\mathbb{R}^2$. [Let $f: M \to \mathbb{R}_+$ be a proper Morse function. For every regular value $\alpha \in f(M)$, the submanifold $M_\alpha = f^{-1}[-\infty,\alpha]$ is a disk with holes. For sufficiently large $\beta > \alpha$, every boundary circle of M_α bounds a disk in M_β. Therefore M is an increasing union of disks: $M = \cup D_i$, $D_i \subset \operatorname{Int} D_{i+1}$. And $D_{i+1} - \operatorname{Int} D_i \approx S^1 \times I$ so a diffeomorphism $M \approx \mathbb{R}^2$ can be built up successively over the D_i.]

4. (a) Let M be an orientable surface of genus p and C_1, C_2 circles in M. Suppose neither circle separates M. Then there is a diffeomorphism $f:(M,C_1) \approx (M,C_2)$. [Represent M as a sphere with handles so that C_i goes around a handle in a standard way.]

(b) Suppose both C_1 and C_2 separate M. In what circumstances is the conclusion of (a) true?

5. (a) Let $M = T - \operatorname{Int} D$ where $D \subset T$ is a disk in a torus. Then M and H_2 are not homeomorphic but $M \times I$ and $H_2 \times I$ are homeomorphic.

(b) The doubles of M and H_2 are diffeomorphic.

6. Two compact oriented surfaces which are diffeomorphic, are diffeomorphic by an orientation preserving diffeomorphism.

7. Every compact nonorientable ∂-surface admits a diffeomorphism that reverses orientation of its boundary.

***8.** Let C, C' be embedded circles in the sphere $S^2 = \partial D^3$. Then every diffeomorphism $f: C \approx C'$ extends to a diffeomorphism of D^3. [C and C' bound disks in S^2 over which f can be extended; etc.]

9. The cobordism groups in dimension 2 are: $\Omega^2 = 0$, $\mathfrak{N}^2 = \mathbb{Z}_2$.

10. If M is an orientable surface of genus p, any $2p + 1$ circles in M which meet each other transversely (or are disjoint) separate M.

11. What are the analogues of Theorem 3.6 and Exercise 10 for nonorientable surfaces?

12. In a compact nonorientable surface every maximal set of disjoint Möbius bands has the same cardinality.

13. Let M be a connected noncompact surface. An *end* of M is an equivalence class $[K,C]$ of pairs (K,C) where $K \subset M$ is compact, C is a component of $M - K$ whose closure is not compact, and $[K_1,C_1] = [K_2,C_2]$ if there exists (K,C) with $K_1 \cup K_2 \subset K$ and $C_1 \cap C_2 \supset C$.

(a) The number of ends is a diffeomorphism invariant

(b) $\mathbb{R}^2$ has 1 end.

(c) For every cardinal number k less than or equal to that of $\mathbb{R}$, there is a surface having exactly k ends.

*(d) A noncompact connected surface M has only a finite number of ends, and has finite connectivity (compare Exercise 17), if and only if there is a compact surface N such that $M \approx N - \partial N$.

***14.** Let M be a connected surface and $X \subset M$ a subset which is the union of an uncountable collection of disjoint circles. Then $M - X$ is disconnected. [Hint: let $\{f_\alpha : S' \hookrightarrow M\} = Y$ be an uncountable collection of embeddings. Prove that $C^1_W(S^1,M)$ is separable and use this to show that Y contains one of its limit points.]

***15.** $\mathbb{R}^3$ does not contain an uncountable collection of disjoint Möbius bands [see hint to Exercise 14].

***16.** Let $f: M \to N$ be a map of degree d between compact connected oriented surfaces without boundary. What relations, if any, exist between the degree of f, the genus of M and the genus of N?

***17.** The *connectivity* $c(M)$ of a compact surface M is the integer $c \geqslant 0$ (if it exists) having the following property. Let $V \subset M$ be the union $V_1 \cup \cdots \cup V_r$ where each V_i is a neat arc, or a circle in $M - \partial M$, and $V_i \pitchfork V_j$ for $i \neq j$. Suppose $M - V$ is connected. Then $r \leqslant c$, and if $r < c$ V is a proper subset of another set V' of the same type. (*Riemann's connectivity* is $c(M) + 1$.)

(a) If ∂M has b components then $c(M) = 2 - \chi(M) - b$.

(b) $c(M)$ is the rank of $H_1(M,\partial M; \mathbb{Z}_2)$.

(c) Express $c(M)$ in terms of b and the genus of M.

Bibliography

Abraham, R.

1. Transversality in manifolds of mappings. *Bull. Amer. Math. Soc. 69* (1963), 470–474.

Abraham, R., and Robbins, J.

1. *Transversal Mappings and Flows*. W. A. Benjamin, Inc., New York, 1967.

Alexandroff, P. S., and Hopf, H.

1. *Topologie I*. Springer-Verlgar, Berlin, 1935.

Artin, M., and Mazur, B.

1. On periodic points. *Annals of Math. 81* (1965), 81–99.

Bochner, S.

1. Analytic mapping of compact Riemann spaces into Euclidean space. *Duke Math. J. 3* (1937), 339–354.

Brouwer, L. E. J.

1. Über Abbildungen von Mannifaltigkeiten. *Math. Ann. 71* (1912), 97–115.

Brown, A. B.

1. Functional dependence. *Trans. Amer. Math. Soc. 38* (1935), 379–394.

Brown, M.

1. The monotone union of open n-cells is an open n-cell. *Proc. Amer. Math. Soc. 12* (1961), 812–814.
2. Locally flat embeddings of topological manifolds. *Annals of Math. 75* (1962), 331–340.

Cerf, J.

1. *Sur les Difféomorphismes de la Sphère de Dimension Trois* ($\Gamma_4 = 0$). Lecture Notes in Mathematics, Vol. 53. Springer–Verlag, New York, 1968.

Dold, A.

1. *Lectures on Algebraic Topology*. Springer–Verlag, New York, 1974.

Dubovickiĭ, A. Ya.

1. On differentiable mappings of an n-dimensional cube into a k-dimensional cube. *Mat. Sbornik N.S. 32* (74) (1953), 443–464 [in Russian].

von Dyck, W.

1. Beitrage zue Analysis Situs I. Ein- and zwei-dimensionale Mannifaltigkeiten. *Math. Ann. 32* (1888), 457–512.

Grauert, H.

1. On Levi's problem and the imbedding of real analytic manifolds. *Annals of Math. 68*, 460–472.

Hadamard, J.

1. Sur quelques applications de l'indice de Kronecker. Appendix to J. Tannery, *Théorie des Fonctions*, Vol. 2, 2nd Edition. Paris, 1910.

Hanner, O.

1. Some theorems on absolute neighborhood retracts. *Arkiv Mat. 1* (1952), 389–408.

Heegard, P.

1. Sur l'Analysis Situs. *Bull. Soc. Math. de France 44* (1916), 161–242. [Translated from: Forstudier til en topologisk teori för de algebraiske Sammenhang. Dissertation, Kopenhagen, 1898.]

Hirsch, M. W.

1. Immersion of manifolds. *Trans. Amer. Math. Soc. 93* (1959), 242–276.
2. On imbedding differentiable manifolds in euclidean space. *Annals of Math. 73* (1961), 566–571.

3. A proof the nonretractability of a cell onto its boundary. *Proc. Amer. Math. Soc. 14* (1963), 364–365.
4. The imbedding of bounding manifolds in euclidean space. *Annals of Math. 74* (1961), 494–497.
5. (with Mazur, B.) *Smoothings of Piecewise Linear Manifolds*. Annals Study 80. Princeton University Press, Princeton, 1974.

Hirsch, M. W., and Smale, S.
1. *Differential Equations, Dynamical Systems and Linear Algebra*. Academic Press, New York, 1974.

Hopf, H.
1. Systeme symmetrischer Bilinearformen and euklidische Modelle der projectiven Raüme. *Viertelgschr. naturfrsch. Ges. Zurich 85* (1940), 165–177.

Hurewicz, W. and Wallman, H.
1. *Dimension Theory*. Princeton University Press, Princeton, 1948.

James, I. M.
1. Euclidean models of projective spaces. *Bull. London Math. Soc. 3* (1971), 257–276.

Jordan, C.
1. Sur les déformations des surfaces. *J. de Math.* (2) *XI* (1866), 105–109.

Kervaire, M.
1. A manifold which does not admit any differentiable structure. *Comm. Math. Helv. 34* (1960), 304–312.

Kneser, H., and Kneser, M.
1. Reell-analytische Strukturen der Alexandroff-Geraden. *Arch. d. Math. 9* (1960), 104–106.

Koch, W., and Puppe, D.
1. Differenzbar Strukturen auf Mannifaltigkeiten ohne abzahlbare Basis. *Arch. d. Math. 19* (1968), 95–102.

Kronecker, L.
1. Über Systeme von Funktionen mehrerer Variabeln. *Monatsberichte Berl. Akad.* (1869), 159–193 and 688–698.

Kuiper, N. H.
1. C^1-equivalence of functions near isolated critical points. In R. D. Anderson, *Symposium on Infinite Dimensional Topology*, Annals Study 69. Princeton University Press, Princeton, 1972.

Lang, S.
1. *Introduction to Differentiable Manifolds*. Interscience, New York, 1962.

Lefschetz, S.
1. *Topology*. American Mathematical Society Colloquim Publications, Vol. 12, New York, 1930.

Lickorish, W. B. R.
1. A representation of orientable combinatorial 3-manifolds. *Annals of Math. 76* (1962), 531–540.

Milnor, J.
1. On manifolds homeomorphic to the 7-sphere. *Annals of Math. 64* (1956), 399–405.
2. *Topology from the Differentiable Viewpoint*. University of Virginia Press, Charlottesville, 1966.
3. *Morse Theory*. Annals Study 51. Princeton University Press, Princeton, 1963.
4. A survey of cobordism theory. *Enseignement Math. 8* (2) (1962), 16–23.

Möbius, A. F.
1. Theorie der elementaren Verwandtschaft. *Leipziger Sitzungsberichte math.-phys.* Classe 15 (1869). [Also in *Werke*, Bd. 2.]

Morrey, C. B.
1. The analytic embedding of abstract real-analytic manifolds. *Annals of Math. 68* (1958), 159–201.

Morse, A. P.

1. The behavior of a function on its critical set. *Annals of Math. 40* (1939), 62–70.

Morse, M.

1. Relations between the critical points of a real analytic function of *n* independent variables. *Trans. Amer. Math. Soc. 27* (1925), 345–396.
2. *The Calculus of Variations in the Large*. American Mathematical Society Colloquim Publications, Vol. 18, New York, 1934.

Munkres, J. R.

1. *Elementary Differential Topology*. Annals Study 54. Princeton University Press, Princeton, 1963.
2. Differentiable isotopies on the 2-sphere. *Mich. Math. J. 7* (1960), 193–197.

Nash, J.

1. Real algebraic manifolds. *Annals of Math. 56* (1962), 405–421.

Palais, R.

1. Morse theory on Hilbert manifolds. *Topology 2* (1963), 299–340.

Poincaré, H.

1. Analysis situs. *J. Ecole Polytechnique, Paris 1* (2) (1895), 1–123. [Also in Oeuvres, t. 6.]
2. Complément à l'analysis situs. *Reindiconti Circolo Matematico, Palermo 13* (1899), 285–343. [Also in Oeuvres, t. 6.]
3. Deuxième complément à l'analysis situs. *Proc. London Math. Soc. 32* (1900), 277–308. [Also in Oeuvres, t. 6.]
4. Cinquième complément à l'analysis situs. *Rendiconti, Circolo Matematico, Palermo 18* (1904), 45–110. [Also in Oeuvres, t. 6.]

Pontryagin, L. S.

1. Smooth manifolds and their applications in homotopy theory. *Amer. Math. Soc. Translations*, Ser. 2, Vol. 11 (1959), 1–114 [translated from *Trudy. Inst. Steklov 45* (1955)].
2. *Foundations of Combinatorial Topology*. Graylock Press, Rochester, 1952.

Riemann, G. F. B.

1. Allgemeine Voraussetzungen and Hülfsmittel fur die Untersuchung von Funktionen unbeschränkt verändlicher Grössen. *J. reine and angewandte Math. 54* (1857), 103–4. [Partly translated in D. E. Smith, *A Source Book in Mathematics*, Vol. 2. Dover Publications, New York, 1959.]
2. Über die Hypothesen welche der Geometrie zugrunde liegen. *Abh. Gesellschaft Wiss. Göttingen 13* (1868), p. 1 *et. seq*. Also in *Werke*, 2nd Edition, p. 272 *et. seq.* [Translated in D. E. Smith, *A Source Book in Mathematics*, Vol. 2. Dover Publications, New York, 1959.]

Sard, A.

1. The measure of the critical points of differentiable maps. *Bull. Amer. Math. Soc. 48* (1942), 883–890.

Schweitzer, P. A.

1. Counterexamples to the Seifert conjecture and opening closed leaves of foliations. *Annals of Math. 100* (1974), 386–400.

Shub, M., and Sullivan, D.

1. A remark on the Lefschetz fixed point formula for differentiable maps. *Topology 13* (1974), 189–191.

Smale, S.

1. Generalized Poincaré's conjecture in dimensions greater than four. *Annals of Math. 74* (1961), 391–406.
2. Diffeomorphisms of the 2-sphere. *Proc. Amer. Math. Soc. 10* (1959), 621–626.
3. Morse theory and a nonlinear generalization of the Dirichlet problem. *Annals of Math. 80* (1964), 382–396.

Steenrod, N.

1. *The Topology of Fibre Bundles*. Princeton University Press, Princeton, 1951.

Takens, F.

1. A note on sufficiency of jets. *Invent. Math. 13* (1971), 225–231.

Thom, R.

1. Quelques propriétés des variétés differentiables. *Comm. Math. Helv. 28* (1954), 17–86.

Whitehead, J. H. C.

1. On C^1 complexes. *Annals of Math. 41* (1940), 809–824.

Whitney, H.

1. A function not constant on a connected set of critical points. *Duke Math. J. 1* (1935), 514–517.
2. Differentiable manifolds. *Annals of Math. 37* (1936), 645–680.
3. On the topology of differentiable manifolds. In R. L. Wilder, *Lectures in Topology*, University of Michigan Press, Ann Arbor, 1941.
4. The selfintersections of a smooth n-manifold in $2n$-space. *Annals of Math. 45* (1944), 220–246.
5. The singularities of a smooth n-manifold in $(2n - 1)$ space. *Annals of Math. 45* (1944), 247–293.
6. Analytic extensions of differentiable functions defined in closed sets. *Trans. Amer. Math. Soc. 36* (1934), 63–89.

Appendix

In this appendix we briefly summarize a few basic facts of analysis and topology.

General Topology

A topological space X is called:

Hausdorff if every pair of points have disjoint neighborhoods;

normal if for every pair of disjoint closed sets A, B there is a continuous map $f: X \to [0,1]$ with $f(A) = 0$ and $f(B) = 1$;

paracompact if every open cover $\mathscr{U} = \{U_\lambda\}_{\lambda \in \Lambda}$ has a locally finite open refinement $\mathscr{V} = \{V_\gamma\}_{\gamma \in \Gamma}$. This means: $\mathscr{V}$ is an open cover of X, each element of $\mathscr{V}$ is contained in some element of $\mathscr{U}$, and for each $x \in X$ the set of $\gamma \in \Gamma$ for which $x \in V_\gamma$ is finite.

The closure of a subset $S \subset X$ is denoted by $\bar{S}$.

A.1. Theorem. *If X is normal and $\mathscr{U} = \{U_\lambda\}_{\lambda \in \Lambda}$ is a locally finite open cover then $\mathscr{U}$ has a shrinking, that is, an open cover $\mathscr{V} = \{V_\lambda\}_{\lambda \in \Lambda}$ such that $\bar{V}_\lambda \subset U_\lambda$.*

A.2. Theorem. *A paracompact Hausdorff space is normal.*

A *partition of unity* subordinate to the open cover $\mathscr{U}$ is a collection $\{f_\lambda\}_{\lambda \in \Lambda}$ of continuous maps $f_\lambda: X \to [0,1]$ having the following two properties: for each $x \in X$ the set of $\lambda \in \Lambda$ for which $f_\lambda(x) > 0$, is finite; and $\sum_{\lambda \in \Lambda} f_\lambda(x) = 1$ for all x.

A.3. Theorem. *A topological space is paracompact if and only if every open cover has a subordinate partition of unity.*

A.4. Theorem. *Every metric space is paracompact.*

A subset A of a space X is *nowhere dense* if its closure $\bar{A}$ contains no nonempty open set; equivalently, $X - \bar{A}$ is dense in X. If X is the union of a countable family of closed nowhere dense subsets X is of the *first category*; otherwise X is of the *second category*.

A.5. Baire Category Theorem. *A complete metric space X is of the second category. Equivalently: the union of any countable collection of closed nowhere dense subsets is nowhere dense, and the intersection of any countable collection of open dense subsets is dense.*

Calculus

Let $U \subset \mathbb{R}^m$ be an open set and $f: U \to \mathbb{R}^n$ a map. A linear map $L: \mathbb{R}^m \to \mathbb{R}^n$ is called *the derivative of f at $x \in U$* if

$$\lim_{h \to 0} |h|^{-1}(f(x+h) - f(x) - Lh) = 0.$$

Here $|h|$ is the norm $\left(\sum_{j=1}^{n} h_j^2\right)^{1/2}$ of the vector $h = (h_1, \ldots, h_n) \in \mathbb{R}^n$. If such an L exists, it is unique and is denoted by Df_x or $Df(x)$.

The map f is called C^1 provided Df_x exists for every $x \in U$ and the map

$$Df: U \mapsto L(\mathbb{R}^m, \mathbb{R}^n),$$
$$x \mapsto Df_x$$

is continuous.

By recursion we define f to be C^r, $2 \leqslant r < \infty$, if the map

$$Df: U \to \mathbb{R}^{mn} = L(\mathbb{R}^m, \mathbb{R}^n)$$

is C^{r-1}. If f is C^r for all r it is called C^∞.

Write $f(x) = (f_1(x), \ldots, f_n(x))$. We call f *(real) analytic*, or C^ω, if in some neighborhood of each point of U, each f^j is equal to the limit of a convergent power series (in m variables). This implies that f is C^∞. We say $\infty \leqslant \omega$.

A.6. Theorem. *f is C^r, $1 \leqslant r < \infty$, if and only if each $f_j: U \to \mathbb{R}$ has continuous partial derivatives of all orders $\leqslant r$.*

Let U and V be open subsets of $\mathbb{R}^n$. A C^r *diffeomorphism* $f: U \to V$ is a C^r homeomorphism $f: U \approx V$ whose inverse is also C^r.

Let $W \subset \mathbb{R}^n$ be open, and $p \in W$. A C^r map $f: W \to \mathbb{R}^n$ is a *local diffeomorphism at p* if there is an open set $U \subset W$ such that $p \in U$ and $f(U)$ is open, and $f|U: U \approx f(U)$ is a C^r diffeomorphism.

A.7. Inverse Function Theorem. *Let $U \subset \mathbb{R}^n$ be an open set and $f: U \to \mathbb{R}^n$ a C^r map, $1 \leqslant r \leqslant \omega$. If $p \in W$ and Df_p is invertible, then f is a C^r local diffeomorphism at p.*

A.8. Implicit Function Theorem (surjective form). *Let $U \subset \mathbb{R}^m$ be an open set and $f: U \to \mathbb{R}^n$ a C^r map, $1 \leqslant r \leqslant \omega$. Let $p \in U$, $f(p) = 0$, and suppose that Df_p is surjective. Then there exists a local diffeomorphism φ of $\mathbb{R}^m$ at 0 such that $\varphi(0) = p$ and*

$$f\varphi(x_1, \ldots, x_m) = (x_1, \ldots, x_n).$$

Proof. After a linear change of coordinates in $\mathbb{R}^m$ we can assume that $\frac{\partial f_i}{\partial x_j}(p) = \delta_{ij}$ for $i = 1, \ldots, n$ and $j = 1, \ldots, m$. Define $h: U \to \mathbb{R}^m$, $h = (h_1, \ldots, h_m)$ where $h_i = f_i$, $i = 1, \ldots, n$ and $h_i(x_1, \ldots, x_m) = x_{i-n}$, $i = n+1, \ldots, m$. Then h is C^r and Dh_p has rank m. By the inverse function theorem h is a C^r local diffeomorphism at p. Therefore in a neighborhood of 0 in $\mathbb{R}^m$, h has a C^r inverse φ. Then $h(\varphi(x)) = x$ for x near 0; this φ satisfies the theorem.

QED

A.9. Implicit Function Theorem (injective version). *Let $U \subset \mathbb{R}^m$ be an open set and $f: U \to \mathbb{R}^n$ a C^r map, $1 \leqslant r \leqslant \omega$. Let $q \in \mathbb{R}^n$ be such that $0 \in f^{-1}(q)$, and suppose that Df_0 is injective. Then there is a local diffeomorphism ψ of $\mathbb{R}^m$ at q such that $\varphi(q) = 0$, and*

$$\psi f(x) = (x_1, \ldots, x_m, 0, \ldots, 0).$$

The proof is "dual" to that of Theorem A.8 and is left as an exercise.

Index

E

F

G

H